CONTROL OF BATCH PROCESSES

CONTROL OF BATCH PROCESSES

CECIL L. SMITH

WILEY

Published by John Wiley & Sons, Inc., Hoboken, New Jersey

Published simultaneously in Canada

For general information on our other products and services or for technical support, please contact our Customer Care Department within the United States at (800) 762-2974, outside the United States at (317) 572-3993 or fax (317) 572-4002.

Wiley also publishes its books in a variety of electronic formats. Some content that appears in print may not be available in electronic formats. For more information about Wiley products, visit our web site at www.wiley.com.

Library of Congress Cataloging-in-Publication Data:

Smith, Cecil L.
   Control of batch processes / Cecil L. Smith.
      pages cm
   Includes index.
   ISBN 978-0-470-38199-1 (hardback)
   1. Chemical process control.   2. Mass production.   I. Title.
   TP155.75.S584 2014
   660'.2815–dc23
                                        2014007295

Printed in the United States of America

10 9 8 7 6 5 4 3 2 1

CONTENTS

PREFACE

This book is based on the premise that the approach to automate a batch process involves three essential components:

Automation equipment. Many alternatives are available, but all are digital. The natural focus is on recipes and sequence control, but neglecting continuous control is a big mistake.

Instrumentation. Resistance temperature detectors (RTDs) for temperatures and coriolis meters for true mass flows address two essential requirements for most batch processes: close temperature control and accurate material charges.

Process modifications. With heat transfer rates varying by 50:1 during a batch, changes such as upgrading a once-through jacket to a recirculating jacket are not just desirable, they are essential to attain good control performance throughout the batch.

Especially when automating a batch process, this book takes the position that the following statements are good advice:

Process modifications are often not just desirable, they are essential. In an early project where process modifications were totally off the table, the project manager correctly observed "we are spending a million dollars to automate a hunk of junk" with a few expletives added for emphasis. Actually, I developed a great respect for this guy. He could express things in ways that the plant manager could understand, which helped save this project from the axe.

Priority must be to achieve a high degree of automatic control. For automating a batch process, much emphasis is placed on recipes, sequence control, and the like. Although definitely important, equally important is close control of key process variables such as reactor temperatures. The sophistication of the continuous control configurations to achieve this in a batch facility rivals their counterparts in continuous plants, especially when coupled with objectives such as minimizing the consumption of a utility such as refrigerated glycol.

Discrete measurements and discrete logic receive little attention in most engineering programs. Consequently, most practicing engineers acquire this knowledge on-the-job in an as-needed manner. One tends to learn what to do, but not always the reasons for doing it that way. Starting with the basics, this book devotes an entire chapter to discrete measurements and discrete logic, the intended audience being process engineers who usually define what the process requires but do not actually implement the logic.

The last part of this book is devoted to recipes, sequence logic, and the like. Especially for highly flexible manufacturing facilities for products such as specialty chemicals, analyzing the requirements of the process and developing a logical organization or structure for meeting these requirements is absolutely essential. But in the end, this is applying well-known concepts (at least within the computer industry) such as structured programming to the logic required to automate a batch manufacturing facility.

This book goes beyond explaining what we should do, but why it is so important to do it in a certain way. Batch automation involves many compromises. Common examples are the compromises made on load cell installations within a process facility. While made for good reasons (especially those pertaining personnel safety), the accuracy of the weight measurement suffers. With regard to batch logic, the main objective is to separate equipment-specific logic from product-specific logic. But process issues often become obstacles to achieving this ideal, necessitating compromises. This book explains the consequences of such compromises.

The basic building block for batch automation seems to be the batch phase or something comparable by another name. Explain this concept to someone with a good understanding of software development. Their reaction: "Obviously, but what's the big deal?" Although normally activated in a manner other than static linking, a batch phase is really the counterpart to a function or subroutine, and these have been around for a long time.

Many requirements of batch automation derive from concepts that have been well known in computer circles for years. Automating a batch production facility involves managing multiple simultaneously active sequences, which entails coordination of two (or more) sequences at certain times, allocating shared resources, avoiding deadlocks, and the like. Multitasking operating systems faced these issues years before they arose in the context of batch automation.

Presentations on batch automation often place great emphasis on various standards. The process industries are favorably disposed to standards, and much effort has been expended developing standards for batch automation. I have not been involved in any way with the standards efforts, but I personally know many of the people who have played key roles. But one side effect of being in computing is that one becomes cynical about certain endeavors that are cherished in certain quarters. For me, standards fall into this category.

Like many people my age, I got into computing the hard way, working with punched cards and learning FORTRAN from IBM's programming manual in 1964. Sort of a rite of initiation, I suppose. But once hooked, computing is so dynamic and fascinating that giving it up is unthinkable. Even after 50 years, I have no regrets.

In the computer industry, change is a way of life. Computer people not only live with change; they expect it! A few years ago, I was working with a company that was using equations for the thermodynamic properties of steam that were 30 years old. The reaction from the computer geeks was enlightening. They wanted to know how anyone could be using the same thing for 30 years. I observed that the thermodynamic properties of steam today are the same as 30 years ago. Being technical people, they understood this and were somewhat embarrassed; they were so accustomed to everything changing that they had given it no thought. But in some respects, they were correct. The thermodynamic properties of steam do not change, but the equations that approximate these properties hopefully improve with time.

From the early days, standards existed in the computer industry, but to what effect? One of the more interesting standards applied to magnetic tapes. The standard essentially stated that the magnetic tape had to be 0.5 in. in width (I have no idea if it was ever updated to metric units). The standard assures that the tape will physically fit into the drive, but presumably, one mounts the tape with the desire to transfer information to or from it. The standard is silent on the latter aspect. Writing tapes in "IBM-compatible format" enabled them to be transferred from one manufacturer's computer system to another.

The computer industry is suspicious of anything with the potential of standing in the way of change. The desire is for the marketplace to determine winners and losers. Consequently, standards do not lead; they follow, often doing little more than crowning the winner. Attempts to drive technology through standards prove futile. Over the years, the computer industry has become very adept at developing standards that fall short of what the name implies.

This book is the fifth I have written on various aspects of process automation. Maybe the sixth should be one on discrete measurements and discrete logic, but written from a process perspective instead of the customary ladder logic and programmable logic controller (PLC) programming perspective. I did a lot of writing while a member of the LSU chemical engineering and computer science faculties from 1966 through 1979. One of my regrets is that I did not continue to write after leaving academia. Only in the last few years did I again take up writing, and it has been a real pleasure.

Fortunately, my wife Charlotte indulges me in such endeavors. If anything, she is too indulging, but I appreciate it more every year.

Houston, TX   CECIL L. SMITH
January 2014

1

INTRODUCTION

Within the span of a half-century, the state of automation in batch facilities has progressed from total manual operation to quite high degrees of automation, even approaching fully automated (or "look ma, no hands").

Prior to 1970, the instrumentation and controls in a batch facility consisted of the following:

Measurement devices. sight glasses for level, pressure gauges using Bourdon tubes, manometers for differential pressure, temperature gauges using bimetallic strips, and so on.

Final control elements. hand valves, start/stop buttons for motors, and so on.

Controller. process operator and possibly a couple of simple loops for temperature, pressure, and so on.

Printed documents called standard operating procedures, working instructions, or similar were provided to the operator for each batch of product to be manufactured. All data were collected manually, often recorded on these printed documents.

Within the industry, batch was commonly disparaged, and not only from those in continuous segments such as oil refining. Those in batch commonly made derogatory comments about themselves such as "We make a batch of

Control of Batch Processes, First Edition. Cecil L. Smith.
© 2014 John Wiley & Sons, Inc. Published 2014 by John Wiley & Sons, Inc.

product and then figure out who we can sell it to": This attests to the high batch-to-batch variability. But as most customers were using these products in processes with similar degrees of automation, the situation was tolerated. "The only reason we have valves is that stopcocks are not manufactured in the size we require": This reflects the influence of chemists in the decision processes within batch facilities in the chemical industry. Often the plant equipment was merely a larger-size version of the equipment used in the laboratory. The author has visited a batch facility that used wooden vessels for reactors, and not that many years ago!

However, one must be careful with such disparaging remarks. The production volumes might be small, but the profit margins are high. Energy costs are usually a relatively minor part of the production costs. The energy crisis in the 1970s devastated the commodity segment of the chemical industry but had only a minor effect on the batch segment. And corporate management took notice.

Realistically, automation of batch processes only became feasible with the advent of digital control technology. Prior to 1970, the control system suppliers largely ignored batch facilities, basically selling them products developed for other markets. This has changed dramatically. All control system suppliers now offer products with features developed specifically for automating batch processes.

Whether intentional or not, the marketing hype could be understood as "just install our equipment and all will be well." Unfortunately, one can spend a million dollars to automate a hunk of junk (a few data points are already available; no more are needed). Install a turbocharger on a Model-T and what do you have? A Model-T. Spending money on controls must go hand in hand with modernizing the production equipment.

This book takes a balanced approach to automating batch production facilities that includes the following topics:

Process measurements. The basis of every automation system is quality data from the process measurements. Chapter 2 focuses on the issues that commonly arise in batch facilities.

Control issues. All batch processes operate over a range of conditions, which imposes requirements on the controls to do likewise. Chapter 3 examines such issues.

Discrete devices. In most batch facilities, the number of discrete values greatly exceeds the number of analog (actually digital) values. Chapter 4 examines the various aspects of discrete measurements and final control elements.

Material transfers. The material transfer systems are generally capable of either transferring one of several raw materials to a destination or transferring a given raw material to one of several destinations.

Chapter 5 examines various equipment configurations for material transfers.

Structured logic. Structured programming and other concepts from computer science are applicable to batch processes, the key being to separate as much as possible equipment-specific logic from product-specific logic. Chapter 6 presents these concepts.

Batch or process unit. Properly defining batch or process units is crucial in any batch automation endeavor. Chapter 7 discusses these issues.

Sequence logic. Batch processing involves conducting the appropriate sequence of activities in the process equipment required to manufacture the desired product. Chapter 8 examines various approaches to implementing sequence logic.

Batches and recipes. In a batch facility in which a variety of products are manufactured in the same production equipment, the product recipe must be the controlling document for manufacturing each product batch. Chapter 9 examines these and related topics.

1.1. CATEGORIES OF PROCESSES

Processes are generally classified as follows:

Continuous. Product flows out of the process on a continuous basis. The conditions within the production equipment are ideally constant, but in practice, minor adjustments to operating conditions are occasionally appropriate.

Batch. Product is manufactured in discrete quantities called batches. The conditions within the production equipment change as the product progresses through the manufacturing process.

Semi-batch. Part of the process is continuous, but other parts are batch. Why not call these semi-continuous?

1.1.1. Continuous Processes

Continuous processes are normally appropriate for manufacturing a large quantity of a given product, such products being referred to as commodity products. The production facility must manufacture the product in the most efficient manner possible, meaning the lowest manufacturing cost.

But there is a downside. Flexibility is sacrificed to gain efficiency. The facility can do what it is designed to do very, very well. Maintaining the efficiency of the production process is essential—commodity markets are competitive, and even those with deep pockets cannot manufacture a commodity product at a loss for very long. But changes in the nature of the raw materials, the specifications for the final product, and so on, that were not anticipated during the process design can lead to significant and expensive changes to the production facility.

1.1.2. Batch Processes

Batch processes are most commonly used to manufacture specialty products such as adhesives, cosmetics, and so on. The lower production volumes do not permit a continuous process to be designed and constructed to manufacture a given product. Instead, several "related" products are manufactured by a given suite of production equipment. In this context, "related" means that the manufacture of these products has something in common, one possibility being that the major raw materials are the same for all of these products.

A few commodity processes are batch, two examples being the polyvinyl chloride (PVC) manufacturing processes and the older style of pulp digesters. But to improve production efficiency, efforts are normally expended to develop continuous processes for commodity products. For example, the Kamyr digester is a continuous digester that has replaced many of the older batch pulp digesters.

Using ethylene as an example of a commodity product and adhesives as an example of a specialty product, there is another important distinction. The ethylene being manufactured today is largely the same as the ethylene manufactured 50 years ago. Perhaps the purity of the ethylene produced today is higher, but basically ethylene is ethylene.

However, the adhesive being manufactured today is very different from the adhesive manufactured 50 years ago. Such products are constantly being tailored to their application, with changes to make it set faster, provide a stronger bond, and so on. Sometimes, tailored products are developed to address a subset of the applications, such as providing a better bond between two specific materials. The number of different adhesives available today is much greater than 50 years ago.

The flexibility of the production facility for a specialty product is more important than its production efficiency. Specialty products tend to evolve, thanks to a development group whose mission is to do exactly that. Many specialty products must perform in a certain manner—adhesives must bond two materials. If an adhesive can be modified to shorten the time required to set, customers view this in a very favorable light, preferring this product over that from competitors and possibly willing to pay a higher price for the product (a shorter setup time makes the customer's production process more efficient). Development groups are tasked with coming up with product enhancements, producing sufficient quantities of the enhanced product for customers to test, and then moving the enhanced product to the production facilities. In many specialty production facilities, predicting what products will be manufactured a year from now is difficult, but rarely is it the same as being manufactured today.

1.1.3. Semi-Batch Processes

A semi-batch production facility uses batch equipment for part of the production process but continuous equipment for the remainder. A common structure for a process is reactions followed by separations. The following approach is possible (and the opposite arrangement is also possible):

1. Use batch reactors for reactions.
2. Use continuous distillation for separations.

The batch reactors discharge into a surge vessel that provides the feed to the continuous stills. On a long-term basis, the throughputs for the two parts of the process must be equal, either by adjusting the frequency of reaction batches, by adjusting the feed rate to the stills, or some combination of the two. On a short-term basis, the throughputs can be different, subject to the constraints imposed by the capacity of the surge vessel.

Many continuous processes utilize batch equipment for certain functions. Oil refineries are normally viewed as the extreme of a continuous process. However, batch operations are commonly used in the regeneration of a catalyst. A production facility with a small degree of batch processing within an otherwise continuous process is not normally deemed to be semi-batch.

1.2. THE INDUSTRY

Flexible batch production facilities are commonly encountered in the specialty chemicals, nonseasonal food processing (coffee, ice cream, etc.), agricultural chemicals (pesticides and herbicides), and similar subsets of the process industries. The companies range from large multinationals to small companies that manufacture products targeted to customers with very specific needs.

1.2.1. Intellectual Property

Go to a paper company and open a discussion on batch pulp digesters. They are generally receptive to having such discussions. Most will admit that what they are doing in the pulp digesters is essentially the same as what their competitors are doing. In some cases, they even visit each other's facilities. They do not view any of their competitive edge being derived from what they are doing in the batch pulp digesters.

With few exceptions, the situation in the specialty batch industry is at the opposite extreme. In some of these facilities, only authorized company employees are allowed to enter. Others will be admitted only when there is a specific need for them to be there, and even then, they are escorted in, they do their job, and they are escorted out. They are only told what they need to know to do their job and are not allowed to just "look around."

There is a valid reason for such approaches. Patent protection has a downside. To be awarded a patent, the filing must include sufficient information so that one "skilled in the art" can reproduce whatever is being patented. To obtain a patent, you have to divulge considerable information relating to your technology. For some companies, this is unacceptable—divulging such information tells your competitors some things that you do not want them to know. This information is a trade secret, which is intellectual property subject to protection. How do you keep a secret? Don't tell anybody! In any trade secret litigation, one must show that extraordinary steps are taken to prevent others from learning about the technology.

This complicates preparing presentations and publications relating to such industries. Anything relating to the process technology has to be expunged. Everyone in the industry likes to tell "war stories." One has to be extremely careful with these, or one can unintentionally divulge information. Even details that seem innocuous to control specialists and the like can be very informative to someone "skilled in the art" and employed by a competitor.

Consequences arise in books like this. Concrete process examples must be avoided. The next section uses the kitchen in one's home as an example of a batch facility. Why? Recipes are widely available in cookbooks. No specialty batch company will allow its recipes to be used as the basis for discussion. Even product names are best avoided, instead using made-up names such as "hexamethylchickenwire." Besides, most of the actual chemical names are so specific to a certain product line that the typical reader would have to look them up anyway. Even worse, a Google search will likely yield little, if anything at all.

1.2.2. Manual Operations

In older facilities, operators provided the process controls. A flow control loop consisted of a vessel with a sight glass, an operator with a clock, and a hand valve. The operator was provided a charge table stating that the vessel level should be x at the start, y after 30 minutes, z after 60 minutes, and so on. The operator adjusted the hand valve so that the flow rate of material out of the vessel resulted in the desired vessel levels. And if you believed what was manually recorded on the production logs, these operators were good! Similar technology was applied to all parts of the batch manufacturing process.

Why did such an approach work? The manufacturing process relied on forgiving chemistry. As a consequence, many of the products could be manufactured in a bathtub outfitted with a good mixer. Just get about the right amount of materials in the tub, agitate it thoroughly, avoid large temperature excursions, and so on, and the resulting product could be sold to someone.

Rarely were products manufactured to order. The plant produced a batch of product, QC performed their analysis to determine the characteristics of the product, and then sales directed this product to the customers that preferred these characteristics. This approach had a downside—periodically, the warehouse had to be purged of product that nobody wanted to buy.

The early attitude was not to automate these processes, but to convert them to continuous processes. But constructing a process to manufacture small amounts of a specific product made no sense economically, especially considering that some of the products had a short life cycle (replaced by a slightly different product with characteristics preferred by the customers). But as increasing energy prices eroded the profitability of many continuous processes, batch facilities became major sources of corporate profits.

1.2.3. Driving Force for Change

With the "if it ain't broke don't fix it" philosophy that prevails in the industry, something is required to drive changes. For batch processes, this something came in the form of an evolution from forgiving chemistry to demanding chemistry. The new products coming out of the development groups were targeted to customers with specific needs, which meant that the products must have specific characteristics.

Tighter tolerances on product characteristics necessitated tighter tolerances during the manufacturing process. Almost overnight, forgiving chemistry was replaced by demanding chemistry. At about the same time, statistical quality control, just-in-time production operations, and the like were recognized as the path forward within specialty batch facilities. Issues such as the following had to be addressed within the production facilities:

1. Tolerances on the total amount of each raw material charged became tighter. Flow measurements consisting of a sight glass and an operator with a stopwatch could not deliver the required performance. Sight glasses were replaced by load cells. Coriolis meters quickly became the preferred flow measurement, in spite of their higher cost.
2. Temperature control became essential. The requirements evolved from "just avoid large temperature excursions" to "control within a few degrees" to "control within 0.5°C." Even "control within 0.1°C" is likely to arise, if not already. Meeting such requirements necessitates that changes such as the following be made:
   (a) Improved temperature measurement (an automation issue).
   (b) More sophisticated temperature control configurations (an automation issue).
   (c) More responsive heat transfer arrangements (a plant equipment issue).
   Enhancements to the controls must be accompanied by enhancements to the plant equipment.

1.2.4. Product Specifications

Those from the continuous industries often find those in the batch industries extremely reluctant to change. Many times this is merely resistance to change, but sometimes there are valid reasons that relate to the nature of the product specifications.

Previously, ethylene was cited as a product produced by a continuous process. One specification is the purity of the product. The purity can be analytically determined using a chromatograph. Occasionally, arguments arise as to whose chromatograph is correct, but once these are resolved, the ethylene either meets the required purity or it does not.

Adhesives were previously cited as products produced by a batch process. Product specifications are developed that state the characteristics that an adhesive must possess to be suitable for a certain application. But in the end, the product must perform properly in the application. If an adhesive is to bond two specific materials, it could conceivably meet all product specifications but not provide the expected bond. This does not arise very often, but one must always remember that the product does not just have to meet specifications; it has to perform in the promised manner. This complicates making changes in a batch facility. Even what seems to be a minor change could potentially have side effects on how the product performs.

The situation is far more serious for a product with a guaranteed life, an example being that the bond provided by an adhesive must remain effective for 10 years. If these bonds deteriorate after 8 years, customers will be more than unhappy—they will start litigating. Liabilities can be huge, potentially risking the financial viability of the company manufacturing the adhesive. In this context, the reluctance to make even simple changes in a batch process are understandable. That even minor changes in how a product is manufactured have the potential to place the company at risk is a very sobering thought.

1.2.5. Automation Technology for Batch

One could get the impression that the control system suppliers were out in front of the curve on this one. Not really. When the need for automation arose within the specialty batch industries, the available equipment for process controls fell into two categories:

Distributed control system (DCS). Although developed by the process instrument companies, oil refining and other large continuous processes were the target market. Not much for batch automation, but in defense of the suppliers, the specialty batch industries were not spending money on automation. This aspect has changed dramatically.

Programmable logic controller (PLC). Developed primarily for the automotive industry, this technology was suitable for manufacturing home appliances, toys, electronic equipment, and so on. PLCs offered robust I/O, could do sequence logic if programmed in relay ladder logic, were very cost-effective, and developed a reputation for high reliability. However, the PLC manufacturers were slow to shed the automotive industry mentality—very understandable for a company whose major customer is General Motors.

More on this in a later section of this chapter.

The know-how for automating a specialty batch facility resided in the chemical companies, not in the control system suppliers. In 1993, the German organization Normenarbeitsgemeinschaft für Meß- und Regeltechnik in der Chemischen Industrie (NAMUR; name translates to "Standards committee for measurement and control in the chemical industry") prepared a document [1] in German whose title translates to "Requirements to be met by systems for recipe-based operations." With input from the major German chemical companies, the document laid out the requirements for control systems suitable for automating specialty batch processes. That such a document was deemed necessary as late as 1993 speaks volumes.

When one tours different specialty batch facilities, one sees a lot of the same equipment. The production companies purchase their glass-lined reactors from the same vessel fabricators. They purchase their chemical pumps from the same pump manufacturers. They purchase their measurement devices and controls from the same suppliers. The difference is not what equipment is being installed, but what is being done within that equipment.

Here, one must distinguish between equipment technology and process technology. With few exceptions, the equipment technology varies little from one company to the next. But what one company is doing within that equipment is far different from what another company is doing. The latter is proprietary, and companies go to great lengths to assure that what is being done within the equipment will not be discussed.

1.2.6. Safety and Process Interlocks

The use of the term "interlock" varies. Some would restrict the term to safety systems and have proposed terms such as "process actions" for similar logic implemented within the process controls. An alternative is use "interlock" to designate the logic required to mitigate the consequences of some event, but to specifically distinguish between the following:

Safety interlock. Interlocks pertaining to personnel protection and to prevent major damage to the process equipment are considered to be highly critical. In most cases, these are relatively simple, but they must be implemented by experienced personnel using the greatest of caution. The design must be based on a detailed analysis of the hazards that could arise. Thereafter, all proposed changes must be thoroughly reviewed.

Modifications to the implementation must confirm that all functions (both those that have changed and those that have not changed) perform as specified.

Process interlock. The purpose of process interlocks is to prevent "making a mess" (such as overflowing a vessel), to prevent minor equipment damage that can be quickly repaired (running a pump dry), and the like. These are normally implemented within the equipment that provides the process control functions. Process interlocks range from simple to extremely complex. Manual overrides are sometimes provided within the logic, but if not, knowledgeable personnel (such as shift supervisors) can usually disable a process interlock. Should production be interrupted for hours due to the failure of a simple limit switch that provides an input to a process interlock, plant management will not be amused. But on the other hand, discipline must be imposed so that the problem that required the override be promptly corrected so the override can be removed.

Herein the terms will be used in this fashion.

An interlock is an implementation of a logic expression whose result is used to force a field device to a specified state, regardless of the desires of the process controls or the process operators. Classifying as a safety interlock or a process interlock has no effect on the discrete logic expression. However, it determines how the interlock will be implemented:

Safety interlocks. Critical interlocks are usually relatively simple, but they must be implemented in highly robust equipment. The options are the following:
1. Hardwired circuits.
2. Programmable electronic equipment dedicated to this function. This typically translates to a PLC. Only discrete I/O is normally required, so a low-end model can be used. This is also consistent with keeping safety equipment as simple as possible.

Process interlocks. These are normally implemented within the process controls, either a DCS or a high-end PLC.

1.2.7. Safe State

The safe state is the state of the final control element on loss of power. To force one or more final control elements to their safe states, a common approach is to remove the power; that is, either disconnect the power supply to electrical equipment or shut off the air supply to pneumatic equipment.

Pneumatic measurement and control equipment has largely disappeared, but with one very important exception. Even today, the actuator for most control valves is pneumatic. The pneumatic diaphragm actuator possesses a characteristic that is very desirable in process applications. On loss of either the control signal or supply air (the power), the actuator for either a control valve or a block valve can be configured in either of the following ways:

Fail-closed. On loss of power, the control valve is fully closed (the actuator is "air-to-open"). The "safe state" for such valves is the closed state.

Fail-open. On loss of power, the control valve is fully open (the actuator is "air-to-close"). The "safe state" for such valves is the open state.

To date, replicating this characteristic in an economical manner has proven elusive for electric actuators for positioning valves. It is possible for solenoid actuators, but even so, many block valves are equipped with pneumatic actuators.

What determines the safe state? From a purely control perspective, there is no basis for favoring fail-closed versus fail-open. The appropriate state for the safe state is determined as part of the analysis of process hazards. On total loss of power, the actuator for the final control element must be configured so that its failure state is the same as the safe state. Once the safe state has been selected, the control logic can be formulated to perform properly.

The logic for safety interlocks must be implemented in such a manner that the final control element is permitted to leave the safe state only if the appropriate conditions are met. If the conditions are not met, the final control element is forced to the safe state, usually by blocking power. But if the conditions are met, the final control element does not necessarily exit the safe state. This only occurs on the following conditions:

1. Either the process controls or the process operators command the device to exit the safe state.
2. Power is applied to the actuator, which means the interlock is not being imposed.

With this approach, safety interlocks and process interlocks differ as follows:

Safety interlocks. Being imposed external to the process controls by disconnecting power from the actuator, the safe state must correspond to the state resulting when all power is removed.

Process interlocks. Being implemented within the process controls, a process interlock imposes a specific state on a field device by forcing the control output to the appropriate value. As the actuator remains powered, a process interlock can impose a state on a field device other than its safe state. The preference is that a process interlock imposes the safe state, but an occasional exception is usually tolerated.

1.2.8. Safety Issues Pertaining to the Product

The previous discussion on interlocks and safe states is from the perspective of equipment safety. In this regard, the issues in batch facilities are largely the same as in continuous facilities.

But in batch processes, another issue arises that is rarely encountered in continuous processes. In the chemical industry, most batch processes utilize reactive chemicals. A variety of situations can arise, examples of which are the following:

1. The reaction proceeds as expected if the order of addition is to add chemical A, then add chemical B, and finally add chemical C.
2. If the order of addition is first A, then C, and finally B, a rapid reaction ensues that overheats and overpressures the reaction vessel.

Interlocks of the type used for equipment protection are far less effective for issues such as this, if even applicable at all.

Assume the reactor has dedicated equipment for feeding each of the chemicals, with individual block valves for each feed. Following the approach for interlocks to address equipment issues, provision could be made to supply power to the block valve for feeding chemical C only when feeding C is acceptable. But who or what turns on the power to the valve? The possibilities include the following:

Process controls. Presumably the process controls would only feed chemical C at the appropriate times. If the process controls cannot be trusted to feed chemical C only at the appropriate times, why trust the process controls to turn on power to the Feed C block valve only at the appropriate times?

Operations personnel. Often this means the shift supervisor or equivalent. The "switch" or its equivalent that supplies power to the feed C block valve could be completely independent of the process controls. The obvious issue is the potential for human error. But in the end, is this a "we could not come up with another way to do it, so we left it to the operators" type of solution?

Another issue that will be discussed in a subsequent chapter pertains to charging the wrong chemical. The intent is to charge B, but D is charged instead, the result being to first add A, then add D, and finally add C. What happens when D is mixed with A, that is, do A and D react? What happens when C is added to the mixture of A and D? Answering these questions raises another question: what is D? There is no way to answer this question. Regardless of what it is, chemical D should not be present in the reacting medium.

Such situations are the results of mistakes. Possibly the truck delivering D was unloaded into the storage tank for B (which means a mixture of B and D was actually added). In the specialty chemicals facilities, the number of chemicals is so large that material is often charged directly from drums. This raises the possibility that one or more drums contained D instead of B. The obvious answer is "we must not allow such things to happen." But exactly how are such situations avoided? Interlocks are not the answer.

And to further drive home how serious this situation can be, sizing relief valves in the chemical industry involves two-phase flow calculations for which the nature of the materials flowing through the relief valve must be known. Adding the wrong chemical has the potential of initiating a runaway reaction that overpressures the vessel. But due to the following, providing adequate protection is not assured:

1. There is no way to know what chemical will be mistakenly charged.
2. There is no way to know what chemicals are flowing through the relief valve.
3. There is no way to be absolutely certain that the relief valve is adequately sized for the situation resulting from the mistake.

This can cause people to lose sleep at night.

1.3. THE ULTIMATE BATCH PROCESS: THE KITCHEN IN YOUR HOME

One measure of a batch process is its flexibility in the range of products that can be manufactured within the facility. By this measure, the kitchen in your home must rank very, very high.

1.3.1. Recipe from a Cookbook

The recipe is the controlling document in a kitchen. Table 1.1 is a recipe for "Mom's Pecan Pie." The terminology is different, but product chemists generate recipes for products with a surprisingly similar structure. There are two main sections:

List of ingredients. Chemists use alternate terms such as **formula** (used herein) or **raw materials**. In either case, the objective is the same: convey what is required to make a batch of the product. Often the chemists will state the formula on the basis of a standard batch size, such as 1000 kg, which can then be scaled to the size of the production equipment.

Directions. Again, chemists use different terms, such as **processing instructions** or **procedure** (used herein). As in the pecan pie recipe in Table 1.1, chemists often itemize the processing instructions as step 1, step 2, and so on. The objective is to state precisely what one must do with the raw materials to manufacture a batch of product.

TABLE 1.1. Example of a Recipe

Mom's Pecan Pie

List of Ingredients

4 eggs	200 gm (1 cup) sugar
240 mL (1 cup) light Karo	4 gm (½ tbsp) flour
1½ gm (¼ tsp) salt	5 mL (1 tsp) vanilla
60 mL (¼ cup) butter	220 gm (2 cups) pecans

Directions

1. Beat eggs until foamy.
2. Add sugar, Karo, flour, salt, and vanilla.
3. Beat well.
4. Stir in melted butter and pecans.
5. Pour into pie crust.
6. Bake 60 minutes at 175°C (350°F).

One of the steps in the recipe in Table 1.1 states, "Mix until foamy." Recipes dating from 50 years ago often contained similar statements. Vessels were often open or outfitted with manholes, permitting operators to visually assess certain aspects pertaining to the batch. Virtually all vessels are now closed, but some nozzles are equipped with sight glasses and perhaps a tank light. However, the trend is to replace qualitative instructions with specific instructions such as "Mix at medium speed for 10 minutes," the objective being to eliminate as much manual intervention as possible.

1.3.2. Home Kitchen versus Commercial Bakery

One observation about recipes for baked goods is in order. The recipe in Table 1.1 is for a home kitchen. Commercial bakeries are very different. Generally, the starting point is a large mixer that combines ingredients for a large number of cakes or pies. These are then formed into individual products and placed on a conveyor belt that passes through the oven. In effect, the result is a semi-batch plant—batch for mixing and continuous for baking. More efficient for making pies but far less flexible than the home kitchen. Baking the Thanksgiving turkey using the equipment in a commercial bakery would not turn out well.

1.4. CATEGORIES OF BATCH PROCESSES

Segmenting the product recipe into a formula and a procedure permits batch processes to be classified as per Table 1.2. The capabilities required of the control system vary greatly with the nature of the batch process, with cyclical batch being the simplest to control and flexible batch the most demanding.

TABLE 1.2. Categories of Batch Processes

Category	Formula	Procedure	Example
Cyclic	Same	Same	Catalyst regeneration for a reformer
Multigrade	Different	Same	Batch pulp digester
Flexible batch	Different	Different	Emulsion reactor

1.4.1. Cyclical Batch

One approach to catalyst regeneration is to provide two catalytic reactors in parallel. While one is in service, the catalyst is being regenerated in the other. Cyclic catalytic reformer units in oil refineries use this approach. While in service the process stream flows through the reactor in one direction. With time, carbon buildup on the catalyst decreases its effective surface area. The catalyst is regenerated by essentially burning the carbon buildup by blowing hot air in the opposite direction.

This is a batch process where a batch consists of one cycle of catalyst regeneration. Logic is required to switch between the two reactors, start the hot air flowing to the one being regenerated, and so on. However, this logic is exactly the same from one batch to the next. The objective is to burn sufficient carbon off of the catalyst to restore its effectiveness, but without unnecessarily degrading the catalyst. A key operating parameter is the temperature of the hot air used to regenerate the catalyst. Usually this is the set point to a temperature controller. A formula in the sense of the list of ingredients in a recipe is not normally used.

For a process to be categorized as cyclical batch, Table 1.2 states the following requirements:

Formula. If a formula is used at all, no changes in operating parameters are normally implemented from batch to batch.

Procedure. Identical logic is used to manufacture every batch.

1.4.2. Multigrade Batch

The older style of batch digesters in the pulp and paper industry is batch. Batch digesters are essentially very large pressure cookers. Each batch or "cook" involves the following:

1. Charging the digester with wood, water, and chemicals
2. Heating the digester to the desired temperature
3. Maintaining the temperature for the specified time
4. Releasing the pressure (or "blowing" the digester) into the recovery system
5. Discharging the contents.

These steps do not vary from one cook to the next. However, selected operating parameters can be changed between batches. One approach is to present a data entry screen to the process operator at the start of each cook. This data entry screen contains parameters such as cook temperature, cook time, chemical-to-wood ratios, and so on, that can be adjusted for purposes such as the following:

1. The water content of the wood is not measured but is not constant. Based on the kappa number (an assessment of the degree of cooking) from previous cooks, the operator adjusts the chemical-to-wood ratios in the formula.
2. The production rate can be increased by increasing the cook temperature and shortening the cook time (cook temperature and cook time are related by a parameter known as the H-factor).

Prior to each cook, the operator is given the opportunity to change the parameters for the next cook. For most cooks, the operator uses the same parameters as for the previous cook. But even so, no cook is initiated until the operator confirms the parameters to be used for the cook.

The PVC process is another example of a multigrade batch process. There are distinct grades of PVC. A set of operating parameters (including quantities such as raw material amounts) are defined for each. The logic is identical (or at least largely the same) for each grade of PVC; only the parameters must be adjusted.

For a process to be categorized as multigrade, Table 1.2 states the following requirements:

Formula. The process differs from batch to batch. For each batch, a set of parameters must be specified. Even though they are often the same from one batch to another, the operator must confirm their values prior to the start of each batch.

Procedure. The logic for this batch is identical to the logic for the previous batch. In this context, "identical" is relaxed to permit minor variations. One approach is to include a logical variable within the parameter set that specifies if a certain action is to be performed. This works well if the possible variations are limited but is cumbersome if the number of possible variations is large.

Some early batch control systems used the term "recipe" to designate what is really only the "formula" in the terminology used today. This worked well for the multigrade category of batch processes but not for the more complex flexible batch category. Today, the term "recipe" is used to encompass both the formula and the procedure.

1.4.3. Flexible Batch

This type of batch facility is commonly encountered in specialty chemicals, with emulsion reactors being a good example. Certainly, it is possible for two successive batches to be for the same product. However, it is also possible that the norm is successive batches of very different products.

This is also complicated by the structure of the facility. For latex paint, three distinct functions are required to produce a batch:

Prepare the emulsion feed for the batch. The proper materials are charged to the emulsion feed vessel, mixed properly to form the emulsion, and then cooled (removing heat from the emulsion feed vessel is easier than removing heat from the reactor).

Carry out the polymerization reaction. Normally, some materials are charged to the reactor, the emulsion is fed (possibly with one or more other feeds) over a period of time (often determined by the rate that heat can be removed from the reactor), and then the reaction is driven to completion.

Perform post-processing. An example is to filter the product.

The nature of the equipment is such that batches overlap. For example, when the emulsion feed to one batch is completed, preparing the emulsion feed for the next batch can be started. Similar possibilities exist between the reactor and the post-processing. As successive batches could be for different products, a recipe (formula and procedure) must be associated with each batch. This is quite analogous to what can occur in a home kitchen—while baking a pecan pie in the oven, the cook can be mixing the ingredients for a cherry pie.

Due to the limitations of the control equipment, some early automation efforts for such processes attempted to develop a formula with logical variables to govern what actions to perform for each batch. The resulting complexity was enormous. For example, suppose materials A, B, and C are required in most batches. The formula certainly includes entries for the amount of each. In some batches, the procedure is to add A, then add B, and finally add C. But in other batches, the procedure is to add A, then add C, and finally add B. As the number of raw materials increases, the possible variations increase rapidly.

There is another serious deficiency to this approach. Suppose for all current batches the procedure is to add A, then add B, and finally add C. Most likely, the logic will be developed specifically for the additions to be in this order (the logic is already complex, so let's not add a capability that is not currently required). But then the development group comes up with a product for which the procedure is to add A, then add C, and finally add B. This product cannot be manufactured until the options are added to both the formula and the logic for the batch.

The complexity of flexible batch processes coupled with the ever-changing mix of products (new ones are added and old ones are discontinued) makes

TABLE 1.3. Automation Functions Required for Batch

1. Safety and process interlocks
2. Basic regulatory control
    (a) Simple PID loops
    (b) Advanced control
3. Discrete device drivers
4. State programmer
5. Sequence logic
6. Recipe management
7. Production control
8. Scheduling

automating such processes very challenging. The only manageable approach is for the recipes to include both the formula and the procedure. That is, the recipe must encompass both parts of a recipe such as in Table 1.1. This requires a structured approach to organizing the logic for the batch. This is the subject of a later chapter.

Control approaches suitable for cyclical batch and multigrade batch cannot be successfully applied to flexible batch. Control approaches appropriate for flexible batch can be successfully applied to cyclical batch and multigrade batch. But especially for cyclical batch, this is overkill.

1.5. AUTOMATION FUNCTIONS REQUIRED FOR BATCH

Table 1.3 lists the wide spectrum of automation functions required to automate batch processes, especially those in the flexible batch category.

1.5.1. Basic Regulatory Control

For pressure control, temperature control, and so on, within a batch facility, simple proportional-integral-derivative (PID) loops are usually adequate. Although most such loops continue to be simple PID, a variety of enhancements are being encountered more frequently:

Cascade control. Cascade is routinely applied to temperature control of vessels equipped with recirculating systems on the heat transfer medium. Flow measurements are more widely installed, which permits temperature-to-flow, pressure-to-flow, and so on, cascades to be installed.

Override control. Generally with the objective of increasing production (which generally means shortening the batch cycle time), batch processes are being operated very close to the pressure limits of vessels, the maximum allowable cooling water return temperature, and so on. Override controls are one way to achieve this mode of operation.

Scheduled tuning. In batch processes, this often means changing the tuning coefficients in a PID controller based on either the time in batch or on the occurrence of some event (such as start or completion of a co-feed).

Model-based controls. In highly exothermic or endothermic reactions, energy balances permit certain variables to be estimated and then used within either the control logic or within a process interlock. But to date, the use of model predictive control (MPC) within batch processes has been limited.

Multivariable control. Temperature is a key variable in almost every batch process. When interaction exists between temperature and some other controlled variable, the controls must be designed to cope with this interaction.

1.5.2. Discrete Device Drivers

The most common final control elements in batch facilities are two-state devices, such as block valves, on/off motors, and so on. A few three-state devices are encountered, such as a two-speed motor with three states: off, slow speed, and fast speed. Most are equipped with one or more contacts (called state feedbacks) that provide discrete inputs to the process controls that confirm that the field device is in a certain state. For a block valve, this results in four possible states:

Closed. The output to the device is calling for the closed state, and the state feedbacks are consistent with the closed state.

Open. The output to the device is calling for the open state, and the state feedbacks are consistent with the open state.

Transition. Following a change in the output to the device, time must be allowed for the state feedbacks to confirm that the device is in the appropriate state. This time is generally referred to as the transition time or travel time.

Invalid. The transition time has elapsed, but the state feedbacks are not consistent with the device state corresponding to the state of the output signal to the device.

The logic to determine the state of the field device from the states of the hardware inputs and outputs is not trivial. The preferred approach is to provide this logic in the form of one or more routines often referred to as discrete device drivers.

1.5.3. Step Programmers

Actions such as starting a raw material transfer in a batch facility usually involve operating more than one discrete device. To start the transfer, one or more block valves must be opened, the transfer pump must be started, and so on. The control logic can drive the devices individually, but alternate approaches offer advantages:

Drum timer. Prior to microprocessor-based controls, a version of the drum timer was used to automate the wash cycle in residential clothes washers. The drum timer consisted of a sequence of steps, each of which specified the desired state of each of the outputs controlled by the timer. The drum timer proceeded from one step to the next based solely on time. Robust versions were occasionally installed in batch facilities, but with digital controls, an enhanced version referred to herein as a "step programmer" is easily implemented. When directed to a specific step, the simpler versions only drive the controlled outputs to their specified states. Separately providing the logic for proceeding to a different step provides capabilities beyond those of the drum timer, such as advancing based on process conditions as well as time, by the ability to skip steps (a jump), to repeat steps (a loop), and so on.

State machine. Developed primarily for manufacturing, a state machine views the manufacturing process as a sequence of process states. A key component of the definition of a process state is the desired states for each of the discrete devices driven by the controls. Upon entry to a process state, the state machine drives each of the controlled devices to the appropriate states. But unlike the step programmer where separate logic initiates the transition from the current step to another step, the state machine also encompasses the logic for proceeding from one state to another.

An essential component of the definition of the step (for the step programmer) or the state (for the state machine) is monitoring for an abnormal condition (a "failure") within the process and reacting to this condition by driving the controlled devices to alternate states. This provides three components to the definition:

1. Desired states of the discrete devices. If the devices do not attain these states, a "failure" is indicated, and an appropriate response is initiated.
2. States of discrete inputs, alarms, and so on. These are to be monitored for a "failure" indication.
3. States for the discrete devices should a "failure" be indicated. Often this alone is an adequate response to the failure, but sometimes additional actions must be initiated.

Developed primarily for manufacturing, state machines often address only those issues pertaining to discrete devices. This likewise applies to most implementations of step programmers. But automating batch processes also involves parameters such as controller set points, target amounts for material transfers, and the like. Extending the formulation of a state machine or step programmer to encompass these is certainly possible.

1.5.4. Sequence Logic

Batch processing usually involves executing an appropriate sequence of actions within the process equipment. The logic to cause this to occur is known as sequence logic. There are several possibilities for implementing sequence logic, all of which will be discussed in more detail in a subsequent chapter:

Relay ladder logic (RLL). Although most appropriate for discrete logic, sequence logic can be implemented as relay ladder logic. Generally, this works well for simple sequences (such as the start-up sequence for vacuum pumps), but not for the more complex sequences generally required for batch processes.

State machine. Logic can be included to determine when to proceed to the next state. However, batch processes do not always execute the states in the sequence in which they are defined.

Sequential function chart (SFC). An advantage of this approach is that parallel sequences are supported as a standard feature.

Procedural language. Most implementations are custom languages derived from either Basic or C, but with the addition of a variety of real-time features. If standard languages such as FORTRAN or C++ are used, the real-time functions must be provided by routines from a function library.

1.5.5. Recipe Management

The simplest situation for batch processing is that product batches cannot overlap. That is, a product batch is initiated (or "opened"), a product recipe is specified, the necessary actions are performed, and the product batch is terminated (or "closed"). Only then can another product batch be opened. For batch pulp digesters, "cooks" are basically executed in this manner (a "cook" constitutes a product batch).

Chemical manufacturing is rarely so simple. The situation can be illustrated for the home kitchen using recipes for pies (as in Table 1.1). Each individual pie corresponds to a product batch. At a point in time, the following is possible:

1. The ingredients for a pecan pie are mixing in a blender.
2. A cherry pie is baking in the oven.

Two product batches are in progress, each using a different product recipe. The objective of recipe management is to make such production operations possible without getting things "mixed up"—like adding an ingredient for a cherry pie while mixing the ingredients for a pecan pie.

PLCs were very effective for automating production operations in automotives, consumer appliances, and so on. Based on experiences in such industries, the suggested approach was to prepare a PLC program to make a pecan pie, a different PLC program to make a cherry pie, and so on. Load the program for the type of pie to be manufactured and proceed to produce one or more pies. Then repeat for the next type of pie. But for the situation illustrated earlier for pie baking, two different pies are being manufactured at a point in time. What program is loaded in the PLC?

Where a large number of product batches using the same product recipe are typically manufactured, this approach could make sense. But on a transition from one type of pie to another, the manufacturing facility must be completely shut down to change the PLC program. Sometimes this is necessary for operations such as equipment cleaning, but not always. Often, "on-the-fly" transitions from one product to another are possible, as is the case for the pie baking example discussed previously. When a complete shutdown is not necessary to make a transition from one product to another, the concept of individual PLC programs for each product is inappropriate. A different approach is required, and one that includes recipe management capabilities.

1.5.6. Production Control

Trying to define what constitutes production control is often a challenge. It differs from plant to plant, from company to company, and so on. Usually, its objective is to make sure that the plant produces the proper products in the proper amounts.

Most corporate information technology (IT) organizations have implemented some form of MRP, which has evolved from material requirements planning (that focused primarily on the raw materials required for production) to manufacturing resource planning (which encompasses all of the resources required by a manufacturing process). In a batch facility, the result is usually a schedule calling for the production of specified amounts of specified products over a specified time interval (such as 1 week).

This must be translated into some number of batches of each product. On one extreme are plants for which the norm is to change the product from one batch to the next (two consecutive batches of the same product occasionally occur, but three batches of the same product are rare). On the other extreme are plants for which the production quantity is translated into a sizable number of batches that may require several days to complete.

In either case, the production quantity stipulated by the schedule is translated into some number of batches using the batch size (which depends on the equipment size) and the expected yield (quantity of product per batch). From time to time, the expected yield will not be realized due to factors such as equipment malfunctions. The norm today is that the production schedule is driven by orders, the objective often being just-in-time (JIT) production.

When the expected yield is not realized, a possible consequence is that one or more customers' orders will not be shipped at the promised time. This can sometimes be avoided by taking corrective action quickly. Usually, this means scheduling one or more additional batches, which can only be done provided the required raw materials and other resources are available. Production control is expected to call attention to such situations, and to provide the necessary data so that good decisions can be made. However, the trade-offs are often complex, and humans usually make the final decision.

1.5.7. Scheduling

As noted earlier, the MRP functions at the corporate level normally generate the production schedule for some period of time in the future for each plant within the company. Sometimes, this is referred to as the long-term schedule.

All plants then translate the production schedule into an appropriate number of batches for each product on the schedule. Although there are exceptions, most plants can execute the batches in whatever order the plant deems most appropriate. And when multiple items of production equipment are installed, the plant usually has the option of selecting the production equipment to be used for each batch. This activity is variously referred to as routing or short-term scheduling.

Routing is an issue only in plants where multiple units of each equipment type are available. This is very common in food processing plants and less common in chemical plants. For example, converting coffee beans into ground coffee products is basically a four-step operation:

1. Roasting the beans, which results in a nonuniform moisture distribution within the beans.
2. Equilibration, which means holding the roasted beans for a sufficient time that the moisture distribution becomes uniform.
3. Grinding.
4. Packaging. Packaging units are usually specific to a particular type of packaging.

The order for the coffee product usually determines the packaging equipment that must be used, but otherwise, the specific items of equipment used for the other steps is at the discretion of the plant.

Suppose our plant has five roasters, eight sets of equilibration vessels, and three grinders. The possible number of paths to produce a given order of coffee is enormous. Furthermore, several coffee orders will be in production at any given instance of time.

Whether all paths are possible depends on the capabilities provided by the material transfer equipment (such as pneumatic conveyors). The typical answer is that most but not all paths are possible. Furthermore, at an instance of time, the conveying equipment is dedicated to making a specific transfer (simultaneous transfers are not usually possible). Some ice cream plants take the options to the extreme. The output from the blending equipment is to vessels on rollers, which can be physically moved to the equipment for freezing the mixture to obtain the final product.

Obtaining an optimum solution to a routing problem is a challenge. And in some plants, one can legitimately ask "why bother." As soon as a solution is obtained, some salesman comes in with an "emergency" order from a major customer that must be produced as soon as possible. Welcome to the real world.

1.5.8. Software Issues

Software development in general and programming in particular have a bad reputation—be prepared for it to take twice as long and cost twice as much. And to get it right, a substantial amount of it must be done a second time. Most early projects in process control experienced similar problems, and thereafter, the desire has been to avoid programming.

However, programming must be understood to extend beyond using procedural languages. Implementing the logic required for automating batch processes as RLL, SFCs, state machines, and other alternatives to procedural languages is essentially programming. Ultimately, the logic is executed by a computer, so similar issues arise regardless of the approach taken.

Regardless of the methodology used to implement the logic, the defects or "bugs" that can arise fall into three categories.

Syntax error. These arise when the rules for implementing the logic have not been followed. The programming tool rejects the specifications and will not attempt to execute the logic. Such bugs will always be detected and can generally be corrected by consulting the appropriate manuals (which are now generally online, so they should be current).

Logic error. These arise when the logic can be executed, but the results are not as expected. Correcting this bug is more complicated, one reason being that it can be the result of either of the following (or perhaps a combination thereof):

1. The logic as specified to the system is not what was intended to be specified. As often stated, computers do what they are told, which is not necessarily what we meant to tell them.
2. The logic is specified as intended, but the user's specifications have one or more flaws.

A second issue is with regard to detecting the presence of the bug. The logic may give the expected results at times, but sometimes it may not. Logic often contains multiple paths, so the defect could be in one or more paths but not in all paths. However, it is possible to devise tests that can be guaranteed to detect errors in the logic. Certain requirements must be met, one of which is that the tests must exercise the logic in all paths.

Timing error. This is sometimes called a "real-time" bug. At a given instance in time, most real-time systems have multiple logic sequences in some stage of progress, and process control systems are no exception. Here is an example from automation of batch processes:

1. The logic for Feed Tank X requires the addition of material A. This logic correctly takes exclusive use of the transfer system for feeding A, executes the transfer, and then releases the transfer system.
2. The logic for Wash Tank Y also requires the addition of material A. However, this logic executes the transfer without first getting exclusive use of the transfer system for feeding A.

When executed individually, both sets of logic perform properly. But if the previously mentioned activities overlap, the results will not be as expected. Depending on how the activities overlap, the results would probably be different, which could complicate locating the problem.

This problem can only arise during the time that Wash Tank Y is executing a transfer. This time interval is often called a "window," and is said to be "open" only between these times. The typical characteristic of a timing bug is that it can only arise when the window is open. If the duration of the transfer to Wash Tank Y is very short and Feed Tank X is the only other user of material A, the logic could perform properly for months before the bug arises. For such bugs that arise infrequently and exhibit different consequences, identifying the root cause of the problem becomes more difficult, one reason being that those who implemented the logic have likely moved on to other projects.

Once the nature of the bug is understood, tests can be devised to reproduce it. But the objective of testing is to identify bugs that are not known. Devising tests that are guaranteed to detect unknown bugs that result from timing is not possible.

To summarize, the consequences on testing are as follows:

Syntax error. No testing is required.

Logic error. Devising tests that are guaranteed to detect these bugs is at least conceptually possible, but for complex applications, extensive testing is required.

Timing error. Devising tests guaranteed to detect these bugs is not possible.

Sometimes the best attitude to take is that "we have bugs; we just do not know where they are."

Timing bugs can arise in process control applications to continuous processes, but they do not seem to occur very often. Examples such as those previously mentioned rarely exist in continuous processes. Being very common in batch processes, the opportunities for defects in the logic become quite numerous.

1.6. AUTOMATION EQUIPMENT

In continuous processes, the emphasis is on analog I/O and continuous control, which is the responsibility of the instrument and control department. The discrete logic required for motor controls, interlocks, and the like is usually the responsibility of the electrical department.

The situation in batch applications is very different, with a much larger emphasis on discrete logic within the process controls. For example, most block valves are outfitted with one or two limit switches to confirm that the valve is in the desired state. Considerable discrete logic is required to support such configurations, leading to the discrete device driver discussed in a subsequent chapter.

Batch operations remained largely manual until the advent of digital technology. The counterpart of a discrete device driver did not exist prior to digital controls. The following describes the traditional approach (analog controls and hardwired logic) followed by the digital technology that replaced it.

1.6.1. Analog

First, a few words about the term "analog" as often used in the industry and as it will be used in this book.

In the 1950s, most of the process automation equipment was pneumatic, with "analog signal" meaning a 3–15 psi pneumatic signal. The 1970s witnessed a steady progression to electronic controls, with "analog signal" meaning a 4–20 mA current flowing in an electrical circuit known as a "current loop." These are true analog signals. PID controllers were either pneumatic or electronic, with the PID logic implemented as analog circuits.

When digital controls were initially introduced, most suppliers chose to continue to use the terminology from conventional or analog controls. The reaction of many within the industry to the transition from analog controls to digital controls was akin to cardiac arrest, so every effort was made to minimize the shock effect of the digital products being introduced. One consequence is that the data within a digital control system are often categorized as follows:

> **Analog values.** Any value within a digital control system that is a counterpart to an analog value in conventional controls is referred to as an "analog value." Although technically digital values, quantities such as

measured inputs and control valve outputs are commonly referred to as "analog values."

Discrete values. Technically, these are two-state values (on–off, true–false, 1–0, etc.). However, discrete values in digital controls are sometimes extended to include more than two states. For example, an agitator with states "Off," "Run Slow," and "Run Fast" is considered to be a discrete device.

With digital control products, all internal processing is performed digitally. The traditional approach for an analog input signal followed these two steps:

1. Digitize the input using an analog-to-digital (A/D) converter to give a raw value.
2. Using digital computations, obtain the "process variable," which is a value in engineering units (°C, kg/min, etc.).

The resulting values are the inputs to the control calculations whose results are the %Open values for the control valves. The digital value for the output was converted using a digital-to-analog (D/A) converter to provide a current loop output to the control valve.

With the introduction of smart transmitters and smart valves, this is changing. As communications technology replaces current loops, smart transmitters provide the input data as digital values in engineering units, and smart valves accept the control outputs as digital values in %Open. In a sense, the A/D conversion has moved from the process controls to the smart transmitters, and the D/A conversion has moved to the smart valves.

1.6.2. Hardwired Logic

Discrete logic is required for motor controls, safety interlocks, process interlocks, simple sequences, and occasionally, discrete control functions. Inputs originate from process switches, including level switches, pressure switches, flow switches, temperature switches, rotary switches, proximity switches, contacts, and so on.

Prior to 1970, discrete logic was implemented via hardwired circuits. Suppose Valve A should open only if Valve B is closed. The safe state for both valves is closed. The process operator commands Valve A to open via push buttons or switches that cause power to be applied to the actuator for Valve A, thereby opening the valve. Valve B is outfitted with a contact on its closed position. If Valve B is fully closed, this contact is also closed. Otherwise, the contact is open. This contact is physically wired in series with the actuator for Valve A. Consequently, power can be applied to the actuator for Valve A only if Valve B is fully closed.

An alternate to hardwired logic is a programmed electronic system, which usually ends up being a PLC. The logic is expressed in the same manner as for hardwired implementations, specifically, as relay ladder diagrams that consist of rungs. Each rung represents a network of series and/or parallel contacts that determine if power can flow from the power source to ground through a coil. The coil is energized only if power flows through the coil to ground. Electricians are taught to troubleshoot problems with the logic by testing for the presence of power at various points in the network.

Using interposing relays and output coils permits quite complex logic to be implemented. Timing can be incorporated by using the following types of output coils:

Delay on. Power must be applied continuously to the output coil for a specified period of time before the output coil energizes.

Delay off. After removing power from the output coil, the coil remains energized for a specified period of time.

Such capabilities permit simple sequences to be implemented as either hardwired logic or within a PLC.

1.6.3. Distributed Control System (DCS)

First appearing in the mid-1970s, these products were specifically designed for large continuous processes such as oil refining. The emphasis was on PID control and related functions required to replace analog control systems, either pneumatic or electronic. Discrete capabilities were supported in the initial DCS products, but not in a manner that permitted extensive use. This improved as the DCS products matured, but even so, more cost-effective alternatives are often required in any application involving extensive discrete logic.

The basic building blocks of a DCS product are the following:

Multifunction controllers. These provide all basic regulatory control functions.

Operator stations. These communicate with the process operators.

These are interconnected by a communications network. The early versions were proprietary, but later versions relied on off-the-shelf technologies such as Ethernet.

To ease the transition from conventional controls to DCS products, the multifunction controllers were configured instead of programmed. The control functions were provided via function blocks, the principal one being a PID controller. To build a simple PID control loop, the PID function block was configured by specifying the source of the process variable (PV) input, the source of the set point (if the inner loop of a cascade), designating options such as controller action, specifying parameters such as the tuning coefficients, and so on. The value of the controller output can be configured as the input to other blocks, and the possibilities include the following:

Valve block. The controller output is retrieved and converted as necessary to provide the output to the final control element.

PID controller. For the inner loop of a cascade, the controller output from the outer loop controller is retrieved and used as the set point for the inner loop.

Characterization function. One approach to implementing split range control is to use characterization functions to convert the controller output to the percentage opening for each control valve.

In conventional controls, the output of one function is connected to the input to another function using physical wires, an approach commonly referred to as "hardwiring." In the DCS products, the connections are specified by software pointers, an approach that is often referred to as "softwiring." Computer graphics technology soon permitted the multifunction controllers to be configured using graphic approaches.

1.6.4. Programmable Logic Controller (PLC)

In 1968, General Motors issued a request for proposals to build a device that eventually became known as a PLC. Designed to replace hardwired logic for controlling stamping machines and the like, the PLC became the first digital technology to successfully compete on a purely cost basis with conventional technology. The early products were limited to discrete I/O, but processed discrete logic rapidly and efficiently. With "programming" based on the familiar (at least to electricians) relay ladder diagrams, the early products relied on a calculator-like input of the program but evolved to directly program relay ladder diagrams using graphical technologies.

PLCs quickly became commodity products with a very attractive "bang for the buck." In addition, the discrete I/O capabilities of these products were both cost-effective and extremely robust, being suitable for harsh environments such as sawmills. In the process industries, the initial PLCs were used largely for implementing interlocks, motor controls, repetitive sequences (such as switching filters), and other functions requiring only discrete logic.

As the PLC suppliers embarked on product enhancements to provide analog I/O and associated continuous control functions, the PLC evolved into a multifunction controller comparable to the multifunction controller in a DCS. In batch applications, the robust and cost-effective discrete I/O continues to be very appealing, with PLCs often incorporated into DCS configurations to provide the discrete I/O.

The term 'PLC' has become a source of confusion, being routinely used to refer to following two very different products:

Low-end PLC. These devices are the preferred technology for implementing motor controls, safety interlocks, and so on, in both continuous and batch processes. These applications require little, if any, analog I/O capabilities and no PID control. These products are used primarily in lieu of traditional hardwired logic.

High-end PLC. These are multifunction controllers that support the various functions required for process control. Networking these PLCs and adding workstations for the operator interface essentially create a PLC-based version of a DCS.

For a batch application, the choices are as follows:

Hardwired logic. Low-end PLC or truly hardwired logic.
Process controls. DCS or high-end PLC.

The primary focus of this book is the process controls for batch applications in the process industries. This book also focuses on the process aspects of the application instead of the systems (bits and bytes) aspects. While the systems aspects differ between DCS and PLC solutions, the process aspects are largely the same. In the end, if an application can be successfully implemented in a DCS, it can also be implemented using PLCs, and vice versa.

REFERENCE

1. Anforderungen an Systeme zur Rezeptfahrweise, NAMUR, NE033 (March 31, 1993).

2

MEASUREMENT CONSIDERATIONS

For any automation endeavor, the process measurements require careful examination. This is "garbage in, garbage out" business—poor data from the measurements lead to poor control performance and decisions that are less than optimal.

Measurements in batch processes fall into two categories:

Online measurements. These are provided by measurement devices installed directly on the process. Regardless of the nature of the product, most batch processes require temperature, flow, pressure, and weight measurements. Measurements required beyond these depend on the nature of the product, and may be pH, density, viscosity, and so on.

Quality control (QC) measurements. These are usually the results from an analytical procedure performed on a sample taken from the process. When these are performed to determine if the product meets specifications, the analyses are performed by trained technicians in the QC laboratory. But when the objective is to provide data on which the process operators can make decisions during the manufacture of the product, the analyses (especially the simple ones) are sometimes performed in or near the control room.

Control of Batch Processes, First Edition. Cecil L. Smith.
© 2014 John Wiley & Sons, Inc. Published 2014 by John Wiley & Sons, Inc.

Attention herein is devoted only to online measurements for temperature, pressure, flow, and weight. Furthermore, only issues of special importance to batch processes will be discussed. An accompanying book [1] in this series discusses the basics of these measurements.

2.1. TEMPERATURE MEASUREMENT

Whenever a reaction is occurring, the temperature of the medium is a critical measurement. The effect of temperature is usually through the rates of the various reactions:

Desired reaction. The higher the rate of this reaction, the shorter the batch cycle and the higher the production.

Competing reactions. This affects the amount of the impurities relative to the amount of the desired product. The following cases arise:
1. The QC lab can analytically measure the impurity levels to determine if the product meets product specifications. The simplest situation is when the specifications impose an upper limit on the composition of the impurity in the final product. However, specifications can be quite complex and involve multiple impurities.
2. The QC lab must measure one or more product properties that are affected by the amount of the various impurities in the product. Product specifications usually state acceptable ranges for such properties.

2.1.1. Resistance Temperature Detectors (RTDs)

The most commonly installed temperature measurement device in batch facilities is the RTD. These are capable of measuring temperature with a resolution of 0.1°C, which is generally sufficient for the majority of batch applications. Today's demanding applications require that temperature be controlled to a tolerance of ±0.5°C. With time, such tolerances tend to become tighter, and should this occur, the current RTD technology may not be up to the task.

The early standards pertaining to RTD technology originated in Europe. Deutsches Institut fur Normung (DIN) is the German industrial standards organization; International Electrotechnical Commission (IEC) is the major international standards organization. The traditional technology for RTDs is a wire-wound platinum element with a resistance of 100 Ω at 0°C that conforms to IEC 60751 "Industrial platinum resistance thermometers and platinum temperature sensors." Previous standards were DIN 43760 and DIN IEC 751, and these are still frequently referenced. Another standard is American Society for Testing and Materials (ASTM) E1137 "Standards Specification for Industrial Platinum Resistance Thermometers."

The wire-wound platinum element is basically handmade, which raises issues pertaining to quality and cost. To reduce the cost, very thin platinum wires are used in constructing RTDs. This has two consequences for batch applications:

1. Shock or vibrations can lead to mechanical damage, typically a break in the thin platinum wire resulting in a total failure of the temperature measurement.
2. Overheating permanently degrades the performance of an RTD. The thin platinum wire deforms but does not break. The result is an inaccurate value for the measured temperature, which negatively impacts the product quality.

In batch applications such as chemical reactors, fermenters, and so on, the RTD element is a minor component of the total cost of a temperature measurement system. Of more concern are issues such as measurement accuracy, maintenance requirements, robustness, and so on.

Most industrial applications of RTDs utilize three lead wires. A bridge network is required to convert the resistance of the RTD element into a voltage that can be sensed by the transmitter. Using three lead wires enables the bridge network to be configured in a way that reduces the effect of changes in lead wire resistance (function of ambient temperature) on the measured temperature. Four-wire installations can further reduce this sensitivity, but installations in industrial processes are rare.

The requirement of a bridge network for each RTD results in two types of installations:

1. Individual temperature transmitters
2. RTD input converter cards within the data acquisition or control equipment.

The temperature transmitter can be located near the measurement point, which shortens the lead wires and further minimizes their effect. However, incorporating distributed or remote I/O technology into the data acquisition or control equipment gives comparable results.

The trend is to convert the input to temperature units (°C or °F with a resolution of 0.1°) within the transmitter or the converter card. A configuration option for the transmitter or converter card determines whether the temperature units are °C or °F. For the converter cards, the configuration option generally applies to all temperatures on the card. Most plants avoid having some temperatures in °C and others in °F, so the ability to individually configure each point is of little value.

With digital transmission between the measurement device and the controls, values in engineering units such as °C or °F are easily transmitted. But as originally developed, converter cards for RTDs could be inserted in conventional I/O racks that also contain A/D converter cards for other inputs from the process. The converter card linearizes the differential voltage from the bridge network to obtain a value in temperature units.

To implement this in a way that is compatible with conventional analog I/O systems, the temperature in engineering units must be expressed as an integer value. The I/O hardware stores the result from a traditional A/D converter in a 16-bit integer storage location. To be compatible, the engineering value for the temperature is also provided as a 16-bit integer value. But since integer values do not have a fractional part, the decimal point is "understood" to be one decimal digit from the right, giving a resolution of 0.1°. An integer value of XXXX is understood to be a temperature of XXX.X°C or XXX.X°F, so 1437 actually means 143.7°C or 143.7°F. With 16-bit integers, the integer value is restricted to the range of −32,768 to +32,767. The corresponding range for temperatures is −3276.8° to +3276.7°, which is adequate for most batch processes.

Microprocessor-based temperature transmitters provide greater flexibility through their extensive sets of configuration parameters. But by accepting multiple inputs (typically 8 or 16), the converter cards have the edge on cost. The two approaches can be mixed, with converter cards used for most RTD inputs but transmitters used to meet special requirements (such as dual inputs with automatic switching in case of the failure of one of the RTDs).

Wire-wound RTDs are widely available in probe assemblies with sizes that fit into standard thermowells. A more recent technology is the thin-film RTD that is constructed by etching a thin film of platinum deposited on a ceramic base. To date, use of this technology in batch facilities has been limited, the issues being as follows:

1. Thin-film RTDs can be mass-produced to give a cost advantage. But since the RTD is a minor component of the total cost of an industrial temperature measurement installation, cost is not a major driving force.
2. Thin-film RTDs are typically 1000 Ω, which reduces the effect of the lead wires. However, three-wire installations using 100 Ω RTDs seem to be satisfactory in batch applications.
3. Thin-film RTDs are less sensitive to shock and vibrations, giving the thin-film RTD an advantage in some batch applications.

Eventually, the thin-film technology will enter batch applications, but the "if it ain't broke, don't fix it" philosophy slows the pace of change.

2.1.2. Thermocouples

The most common application for thermocouples is for combustion processes where the process temperatures exceed the upper temperature limit for RTDs.

Heat treating of metal parts is a batch process in which such temperatures can be encountered. However, these are not reaction processes, so the lower accuracy (approximately 1% of reading) of thermocouples is usually acceptable. Better accuracy can be realized by custom calibrating each thermocouple measurement, but the additional expense can rarely be justified in batch applications.

Another advantage of thermocouples over RTDs is that a significant number of thermocouples can be easily multiplexed so that a single readout device can be used. An example of such an application is a cold room where the temperature must be measured and recorded at several locations within the room. Such data are required to confirm that the product stored in the cold room does not exceed its allowable temperature. The required measurement accuracy is easily provided by thermocouples. The rate that each thermocouple must be scanned is also quite slow relative to the capabilities of current electronic equipment.

2.1.3. Thermistors

The technology installed in batch processes is largely driven by process requirements. Today, measuring temperature with a resolution of 0.1°C is normally adequate. The installation of RTDs instead of thermocouples is largely driven by this requirement. Will a better resolution be required in the future, and if so, what technology must be installed?

Where such a requirement arises, one option is to install a thermistor, which can be thought of as a nonmetallic RTD. The resistance of a thermistor is normally in the megohms, and furthermore, it has a large (and negative) sensitivity of resistance to temperature.

Thermistors are widely used within equipment, such as a temperature-controlled oil bath, that is routinely installed in the process industries. But to date, thermistors are rarely installed for process temperature measurement. The temperature range is more restricted than for RTDs but easily covers the required temperature range for many important processes (for water-based processes such as fermentations, a range of 0°C–100°C is usually sufficient).

Should a requirement to measure temperature with a resolution of 0.01°C arise, a thermistor is one option. But given sufficient incentive, other approaches will no doubt be explored.

2.1.4. Thermowells

The installation of bare probes is acceptable only in applications such as measuring the temperature of nonhazardous gases at low pressure, one example being temperatures in air ducts. Rarely can bare probes be used in industrial batch processes. Instead, the temperature probe must be installed within a thermowell that provides the necessary protection from the process fluids. The thermowell also permits the probe to be accessed without interrupting the process, a definite advantage for a continuous process but less so for a batch process that is basically shut down between the end of one batch and the start of another.

The temperature probe senses the temperature at a point inside the thermowell near the tip. The usual practice is to refer to this temperature as the process temperature. In reality, the relationship between the temperature of the process fluid and the temperature of this point near the tip of the thermowell depends on the structure of the thermowell, the characteristics of the process fluid, and so on. This relationship is neither simple nor easily quantified.

The nature of the process fluid greatly complicates the thermowell selection and installation. Some thermowells must be fabricated from expensive metals (tantalum), some must be coated with corrosion-resistant materials such as glass, and so on. Inserting the thermowell through a nozzle at the top of the vessel usually raises serious structural issues, so alternatives such as inserting through a bottom or side nozzle may be necessary despite issues such as leaks. Sometimes the temperature probe is essentially incorporated into vessel internals such as baffles whose real purpose is to enhance mixing.

For a thermowell inserted into a vessel, the force resulting from the fluid flowing across the thermowell is substantial. Structural issues favor a short thermowell. However, most reactors are outfitted with a jacket for either heating or cooling. Heat can flow from the jacket to the metal in the reactor walls to the thermowell (or vice versa), causing the temperature probe inserted into the thermowell to read high (or low). This effect is greater for a short thermowell than a long one. But if installing a longer thermowell entails thicker walls to obtain the necessary structural rigidity, the advantages of a longer thermowell are reduced.

Calibration often means removing the temperature probe and comparing its reading to that of the calibration equipment. In no way does this assess the contribution of the thermowell. Furthermore, assessing the contribution of the thermowell under reacting conditions is rarely practical. Between batches, one could fill the reactor with water (or a fluid that is a liquid at reacting temperatures), insert a high-accuracy probe directly into the fluid within the vessel, and compare the temperature sensed by the two probes. But in most cases, the reacting medium is a fluid that is very different from the inert fluid used during calibration. This affects the error in the temperature measurement, and this error likely varies as conditions change during the batch. Given such considerations, calibrations of this type are rarely undertaken.

Generally, the more protection that must be provided to the temperature probe, the greater the deficiencies of the probe. These include impaired accuracy, slower response time, and so on. Unfortunately, the alternatives are limited. Those critical of a temperature measurement installation are always free to propose something better.

2.1.5. Accuracy versus Repeatability

The previous discussion indicated that an RTD can measure the process temperature with a resolution of 0.1°C. This can be understood from two different aspects:

1. The smallest change in process temperature that can be sensed is 0.1°C.
2. The process temperature must change by at least 0.1°C to assure that the measurement device responds. For smaller changes, the measurement device may or may not respond, depending on the temperature relative to the quantization intervals.

A measurement device with a resolution of 0.1°C does not necessarily provide an accuracy of 0.1°C. In most industrial reactor applications, the temperature affects various product characteristics, which must be assessed within the QC lab to assure that the product meets specifications. However, the product specifications rarely explicitly state the reactor temperature. Instead, specifications state factors such as maximum impurity levels, allowable range for product density, and so on. For such factors, the accuracy of the QC lab measurement is of paramount importance.

What about the accuracy of the temperature measurement? Fortunately, repeatability is usually more important than accuracy. This is where those very familiar with the process technology play a crucial role. During commissioning of a new product within a batch facility, initial values for targets for parameters such as reactor temperature are available from the development group, from a sister plant, or otherwise. But viewing these as "cast in concrete" may not be wise.

Suppose the initial value for the target for the reactor temperature came from the development group. Is their reactor identical to the reactor in the production facility? In other words, is the size the same, is the mixing identical, are the temperature probes installed at the same locations, and so on? The answer to one or more of these is likely to be "not exactly." If so, is the reactor temperature recommended by the development group appropriate for the production reactor? Maybe not exactly.

In a sense, if a product meets all of the criteria stated in the product specifications, the product is acceptable, regardless of the temperature within the reactor. What if the product does not quite meet all specifications? Based on the results of the various analyses performed by QC, those familiar with the process technology can suggest various changes that should give the desired results. One possible change is a small adjustment in the reactor temperature. Maybe this is compensating for an error in the temperature measurement in the reactor used by the development group, maybe this is compensating for an error in the temperature measurement in the production reactor, or maybe it is compensating for something entirely different, such as the degree of mixing. The initial objective is always to make whatever change is required to make a salable product. If only small changes are required, rarely are the reasons pursued. But for large changes, the cause should be identified and corrective actions taken. What is considered "small" versus "large" is usually at the discretion of those familiar with the process.

In the ideal world, all production reactors would perform in exactly the same manner, and thus, exactly the same temperature would be used for all. Efforts should be expended to minimize the differences, but completely eliminating them is unrealistic. Process equipment purchased at different times should be similar but not always identical. After all, equipment manufacturers are expected to incorporate the latest technology into their products, otherwise their products become "long in the tooth." As the reaction equipment is "not exactly the same," provision must be made for the reaction temperature to be "not exactly the same."

Those knowledgeable in the process technology are "fine-tuning" or "trimming" operating variables, such as the reactor temperature, with the objective that the process makes a salable product, operates more efficiently in some respect, and so on. For this to be possible, small changes in parameters such as the reactor temperature must be allowed. To those (including control personnel) who are not familiar with the process technology, this resembles magic, or perhaps the "laying on of hands." But given the results, one must be careful imposing anything that would preclude such activities.

2.1.6. Multiple Probes

Failures of temperature probes in batch facilities are infrequent occurrences. Probe assemblies are available that are fabricated with two RTDs and cost only nominally more than a probe with only one RTD. Temperature transmitters are available that accept two RTD inputs, and some will argue for purchasing such a transmitter and connecting the second RTD. However, most seem content with merely having a spare RTD readily available so it can be connected quickly if needed. The objective seems to be simplicity as per the "keep it simple stupid" philosophy.

Batch reactors are generally equipped with as good an agitator as practical. The usual objective is for the reaction medium to be perfectly mixed, which means uniform conditions throughout the reaction medium. Nonuniform distributions in composition, temperature, and so on, within the reactor affect the reactions that occur within the reactor and consequently the product characteristics. But for a reaction medium with a high viscosity, achieving perfectly mixed is not realistic. And in some cases, excessive agitation can adversely affect the reaction medium (e.g., excessive agitation can break an emulsion).

Multiple temperature probes are installed in many reactors. A common situation is where material is initially charged to provide a "heel" of nonreactive material, the objective being to sufficiently cover the agitator so that it can be started. With no agitation, there is essentially no heat transfer. Although a temperature probe is normally inserted so that the heel temperature can be measured, the measurement is meaningless until the agitator is running.

As additional materials are charged to the reactor, the level in the reactor increases. Questions arise as to whether or not the lower temperature probe (installed to measure the temperature of the heel) is representative of the temperature in the main portion of the reacting medium. The concerns are addressed by installing one or more additional temperature probes. The differences in the temperatures measured by the probes reflect the nonuniformity of the conditions within the reaction medium.

The multiple probes cannot be viewed in the customary sense as redundant measurements. In a power plant, temperature measurements that are used for control purposes are likely to be triple-redundant. That is, three sensors are installed in such a fashion that their measured values should be identical. But even calibrated thermocouples sensing temperatures in the 1000°C range will not give identical values. Various algorithms have been proposed to compute a "believed value" from these multiple inputs.

In batch applications with multiple temperature probes, the various temperatures are not expected to be identical, so using such algorithms is not normally appropriate. But with multiple probes, what is meant by "reactor temperature"? There are several possibilities:

1. The probe that is designated as the "reactor temperature" changes as the batch progresses. Initially, the heel temperature indicated by the lower probe is the only option (no other probe is submerged in the reacting medium). But as other probes are adequately submerged, their values could be used. Sometimes the decision as to which probe to use is incorporated into the reaction logic; sometimes the decision is up to the process operator.

2. The "reactor temperature" is an average of the values from the submerged probes. This could be a simple arithmetic average, or it could be a weighted average. In practice, this approach does not seem to be common.

3. The "reactor temperature" is the maximum or minimum of the values from the submerged probes. For example, in highly exothermic reactions where the temperature could increase rapidly, a maximum could be imposed on the temperatures throughout the reaction medium.

2.2. PRESSURE MEASUREMENT

With regard to sensing pressure, the following possibilities arise in batch processes:

1. Atmospheric. Pressure is slightly above or slightly below atmospheric pressure.

2. Pressurized. Pressure is significantly greater than atmospheric pressure.
3. Vacuum. Absolute pressure is near zero.

Sensing pressures well above ambient in a batch process is rarely different from sensing such a pressure in a continuous process, so sensing such pressures will not be discussed herein. However, issues pertaining to pressure tests to detect leaks will be discussed at the conclusion of this section.

2.2.1. Atmospheric

This mode of operation is common for batch distillation columns, including reactors for which one of the products is withdrawn via a separation column. The objective is to operate the process in either of the following:

Just above atmospheric pressure. No oxygen leaks into the process, but materials within the process can escape to the environment. A shutdown is triggered should the pressure fall below atmospheric.

Just below atmospheric pressure. Materials within the process cannot escape to the environment, but oxygen leaks into the process. A shutdown is triggered should the pressure rise above atmospheric.

To operate in this manner requires two capabilities:

Vent. As the pressure rises, more gas is released into the vent system. Usually the pressure within the vent system is below atmospheric pressure. Suction is provided by a blower, not by a vacuum equipment.

Inert gas. As the pressure decreases, an inert gas is bled into the process. The inert gas can be anything that does not affect the process. Nitrogen is sometimes required, but whenever possible, a less expensive inert gas such as methane, carbon dioxide, or, occasionally, steam will be used.

Control options include split range or two controllers with separated set points [2].

Herein attention is directed to measuring the pressure within the process. The pressure differential between process and atmospheric is very small and is often measured in units such as cm H_2O. A narrow span (such as 20-cm H_2O) is required.

This application can be affected by variations in the atmospheric pressure, so the pressure must be measured as gauge pressure (the difference between process pressure and atmospheric pressure). Two approaches are encountered:

1. Install a differential pressure transmitter with one side open to the atmosphere. Prior to automation, manometers were used for this purpose, and the counterpart to a manometer is a differential pressure transmitter.
2. Install a gauge pressure transmitter with a narrow span.

With smart pressure transmitters, the two approaches are equivalent—the gauge pressure transmitter and the differential pressure transmitter have the same internal pressure-sensitive components.

2.2.2. Vacuum

The batch processing occurs only at low pressures, but batch operations also involve start-up and shutdown. In some cases, the vacuum can be established as rapidly as possible. But sometimes this is not advisable.

Whenever vacuum must be established in a careful manner, the vessel must be equipped with two pressure measurements:

Wide span. A measurement range of 0–800 mmHg is common.

Narrow span. The measurement range depends on the pressure required for the batch processing, but ranges such as 0–50 mmHg are encountered.

The wide span measurement is used for start-up (and possibly for shutdown); the narrow span measurement is used to control the vacuum at the target required for batch operations.

The accuracy of a pressure measurement is normally expressed as a percentage of the upper range value. Most modern pressure measurements are capable of higher accuracy, but to keep the math simple, an accuracy of 0.1% of the upper range value will be used. In the engineering units of the pressure, the accuracies are as follows:

Measurement Range	Accuracy
0–800 mmHg	0.8 mmHg
0–50 mmHg	0.05 mmHg

At high vacuum, a change in pressure of 0.8 mmHg is significant, being 1.6% of the upper range value of the narrow span measurement.

A similar situation arises with regard to resolution. When the interface to the pressure transmitter is an analog signal, the typical converter provides a resolution of 1 part in 4000. An input with a measurement range of 0–800 mmHg can be converted with a resolution of 0.2 mmHg. This is adequate for start-up, but not for control at high vacuum. An input with a measurement range of 0–50 mmHg can be converted with a resolution of 0.0125 mmHg.

With a digital interface to the pressure transmitter, the resolution issues are similar. For a measurement range of 0–800 mmHg, the input value could be represented to 0.1 mmHg. That is, the format of the value is XXX.X in engineering units of millimeter of mercury. But for a measurement range of 0–50 mmHg, the value could be represented to 0.01 mmHg, the format being XX.XX in engineering units of millimeter of mercury. For closed-loop control at high vacuum, the latter must be used.

The desire is to use the narrow span pressure transmitter when the pressure is within its measurement range; otherwise, use the wide span pressure transmitter. One way to accomplish this is as follows:

1. Configure a point for the wide span pressure transmitter.
2. Configure a point for the narrow span pressure transmitter.
3. For the vessel pressure configure a derived or composed point that contains the following logic based on the value from the narrow span pressure transmitter:

   Within range. Use the value from the narrow span pressure transmitter as the vessel pressure.

   Not within range. Use the value from the wide span pressure transmitter as the vessel pressure.

With few exceptions, the remaining control logic can use the vessel pressure without concern for its source.

2.2.3. Establishing Vacuum

In most cases, a vessel can be evacuated as rapidly as possible—fully open the valve to the vacuum system, run the vacuum pump at maximum speed, or whatever. But occasionally, consequences can potentially arise, an example being as follows:

1. Material in the vessel contains a volatile component.
2. As the pressure is lowered, the volatile component begins to vaporize.
3. "Boiling off" the volatile component generates a foam layer on the surface.
4. If excessive, the foam enters the vacuum system, usually with adverse consequences.

The desire is to establish the vacuum as rapidly possible, but without generating an excessive amount of foam.

If vacuum is established as rapidly as possible, the pressure behaves as follows:

1. Pressure drops rapidly until the volatile component starts to vaporize.
2. While the component is vaporizing, the pressure drops slowly.
3. After most of the component has been vaporized, the pressure drops rapidly again.

The potential for foaming arises in the second step (volatile component is vaporizing). For materials with a tendency to foam, the volatile component must be removed more slowly so that the foam layer on the liquid surface does not build up to the point that the foam enters the vacuum system.

Since the culprit is the foam, sensing the location of the top of the foam layer would be very useful. Very few possibilities are available, and installing such sensors in the vessels commonly encountered in batch applications is usually impractical.

2.2.4. Flow to Vacuum System

When introducing automation into an older facility, one option is to inquire as to how the process operators went about establishing vacuum in the vessel without introducing foam into the vacuum system. Using a hand valve, the procedure went as follows:

1. Start with the valve fully closed.
2. Slowly open the valve, listening to the "whistle" of gases flowing through the valve. With experience, the operators learned the appropriate pitch of the sound that could be tolerated without excessive foaming.
3. As the pitch changed with decreasing flow through the valve, the process operator would increase the valve opening.

This continued until the valve was fully open.

Basically, the operators are using the hand valve as a crude sonic flow meter. Computers do not do this very well (this may change in the future). Essentially, the operator is attempting to open the valve so as to maintain a constant flow into the vacuum system. For the controls to use such an approach, a flow meter must be installed in the vacuum piping. This is impractical, especially for the off-gases typically encountered in batch applications.

With no flow measurement, the process controls must rely solely on the measured value of the vessel pressure. A pressure indication is available to the process operators, but they do not rely on this measurement when establishing vacuum. One normally assumes that automatic approaches will provide performance superior to that of the operators. But for this application, this is not assured. When the potential consequence is foam entering the vacuum system, one tends to take a conservative approach.

2.2.5. Pressure as a Function of Time

Control logic could be implemented to adjust the vacuum valve opening such that the pressure in the vessel drops at a specified rate. However, a constant rate is not appropriate—the pressure decreases slowly while the volatiles are being removed.

With digital systems, a more complex relationship between pressure and time could be used, an example being the following:

Prior to vaporization of the volatile component. Starting at atmospheric pressure, a specified rate of change of the pressure must be maintained until a specified pressure is attained. The pressure for terminating this ramp is the pressure at which significant amounts of the volatile component begin to vaporize.

Vaporization of the volatile component. A slower rate of change of the pressure is maintained until another specified pressure is attained. The pressure for terminating this ramp is the pressure at which most of the volatile component has been vaporized.

Subsequent to vaporization of the volatile component. A higher rate of change of the pressure is maintained until the target is attained for the pressure.

This is a version of logic referred to as "ramp and soak" that is commonly used in heat treating furnaces. Such logic is configured in segments, each consisting of two actions:

Ramp. Change the controlled variable (in this case, pressure) at a specified rate until a specified value (the target) is attained.

Soak. Maintain the controlled variable at the target for a specified time.

For each segment, three coefficients are required: ramp rate, target, and soak time. For establishing vacuum, no soak is required, so the soak time is zero for all segments. Assuming the target for the final ramp is separately specified, five coefficients must be specified—three ramp rates and two targets—for the logic to establish the vacuum.

As normally implemented, the ramp and soak function provides the set point to a pressure controller that must be properly tuned. The ramp and soak functions normally offered by digital process controls provide features such as stopping the ramp should the difference between the target pressure and the actual pressure exceed a specified value. For this example, the major concern is avoiding vaporization rates that cause excessive foaming. With no flow measurement, about all one can do is to impose a limit on the valve opening during each segment of the ramp and soak. However, the customary implementations of ramp and soak logic do not provide such a feature.

One possibility for obtaining a starting point for the ramp rates and targets is to have the process operators establish vacuum manually and record the pressure as a function of time (a trend plot of the pressure). One could then approximate this trend with three or more ramp segments. Adding ramp segments is not a problem for digital process controls. However, someone must provide the ramp rate and target for each ramp segment. The latter proves to be the major obstacle. Process controls permit the values of the ramp rate and target for each segment to be easily adjusted, but this raises two questions:

1. Who will do it?
2. On what basis will they do it? And be more specific than simply "based on operational experience."

For establishing vacuum, the only operational data from previous batches is the pressure profile. One normally starts with very conservative ramp rates, making excessive foaming unlikely. The consequence is that establishing vacuum requires considerable time, which will not go unnoticed. Someone will have to increase ramp rates, and that someone may be you.

In furnaces for heat treating metal parts, the metallurgists specify a temperature profile in terms of the ramp rates, targets, and soak times for the furnace temperatures. Normally, the basis for these values is to heat the metal part to the desired temperature without incurring excessive thermal gradients within the part that would result in damages such as warping or cracking. For establishing vacuum without excessive foaming, there is no corresponding basis for suggesting the pressure profile.

2.2.6. Valve Opening as a Function of Pressure

Providing pressure control while establishing vacuum is not mandatory. Even though a pressure indication is available to the process operators while establishing vacuum, they rely largely on their crude sonic flow meter.

One possibility is to mimic the operator but using the vessel pressure instead of the audible feedback. The following sequence could be pursued:

1. Open the control valve to a specified value.
2. Wait until the pressure attains a specified value.
3. Repeat until either the target for the vacuum is attained or the control valve is fully open.

One opens the valve to a specified value, waits until the pressure attains a specified value, increases the valve opening by a specified amount, waits until the pressure attains another specified value, and so on. A starting point could be obtained by having the operators establish vacuum manually and noting the pressures at which they change the valve opening. Unfortunately, translating hand valve openings to control valve openings cannot usually be done accurately.

The basis for implementing such logic is a table specifying the valve opening as a function of vessel pressure. The following is an example of such a table:

Pressure	Valve Opening
Initial	M_0
P_1	M_1
P_2	M_2
.	.
.	.
.	.
P_n	M_n

Start establishing vacuum by opening the valve to M_0. When the pressure attains P_1, open the valve to M_1; when the pressure attains P_2, open the valve to M_2, and so on. Usually the value for M_n is 100% — that is, the sequence is continued until the valve is fully open. Values for M_0, M_1, and so on, and P_1, P_2, and so on, are easily adjusted, but again, on what basis? The higher the variability of the composition of the charged material, the more conservative these parameters must be set.

The value for M_0 is often surprisingly large. Most valves are equal percentage. For accurate values, the inherent characteristics of the valve must be known. But typically, a 30% opening of an equal percentage valve gives a $\%C_V$ of about 5%; a 50% opening gives a $\%C_V$ of slightly over 10%.

While establishing vacuum in the vessel, precise pressure control is not necessary. As long as no foam enters the vacuum system, any pressure profile is satisfactory. Which is preferable — pressure control with a preprogrammed pressure set point or no pressure control with a preprogrammed valve opening? The latter is usually viewed as the simplest and consequently most appealing.

2.2.7. Leaking Agitator Seal

When vessels must be pressurized using known carcinogens such as vinyl chloride, leaks must be taken very seriously. As vessels are pressurized and depressurized during batch operations, the potential for leaks from mechanical equipment such as agitator seals must be considered.

Usually, the primary concern is the exposure of personnel to health hazards. Environmental monitors are usually installed to detect such situations, and this is the final defense against the hazard. The issue becomes what steps can be taken within the batch logic to avoid releases that would set off the environmental monitors.

Prior to feeding potentially hazardous chemicals to a vessel, a pressure test can be conducted, the typical procedure being as follows:

1. Close all valves, vents, and so on.
2. Pressurize the vessel with an appropriate gas (air, nitrogen, carbon dioxide, natural gas, etc.), and record the initial pressure.
3. Wait a specified period of time.

4. Check the vessel pressure, and alert operations personnel if the change exceeds a specified value.

This procedure applies to pressurized vessels; a slightly modified procedure can be applied in vacuum applications.

The viability of this approach depends on three factors:

The size of the vessel. The larger the vessel, the longer the wait time required to detect the change in pressure. For vessels such as reactors, an equipment utilization cost is computed as the $ per hour of vessel utilization. For large vessels, the $ per hour is also large, so increasing the wait time becomes costly.

Pressure measurement. Modern pressure sensors are both very accurate and very repeatable (this application relies on the repeatability of the measurement, not its accuracy). But for a fixed quantity of gas within the vessel, the vessel pressure also depends on the temperature. The wait time must be sufficiently long that the loss in pressure from leaks is larger than the change in pressure that could result from temperature issues. Temperature compensation is possible, but temperature probes installed to sense liquid temperatures in agitated vessels are far different from probes designed to sense gas temperatures.

Magnitude of the leak. The smaller the leak to be detected, the longer the wait time.

Leaks around agitator seals tend to be small. For a large vessel, the required wait time will become a major component of the time required to produce the batch. This translates into an unacceptable reduction in productivity.

One advantage of the previously mentioned approach is that it detects any leak—that is, it is not specific to a given source of the leak. But when a leak from a specific source is of concern, alternative approaches should be explored.

For pressurized vessels, a traditional approach to detecting leaks through the agitator seals is a simple procedure involving soap suds—primitive yes, but very effective. Total automation is the ideal, but in many batch applications, completely eliminating manual activities is impractical.

2.3. WEIGHT AND LEVEL

Both weight and level are measures of the contents of a vessel. Weight is a mass; level is a distance. In batch processes, weight is usually of more significance than level. Following a few observations regarding level, attention will be directed entirely to weight.

2.3.1. Level

Level and weight are related. For vessels with a simple geometry such as a cylindrical-shaped vertical tank with flat ends, the weight can be expressed in terms of level very simply:

$$W = \rho A H,$$

where

A = Cross-sectional area
H = Height of liquid measured from the bottom of the tank
ρ = Liquid density
W = Weight of liquid in the tank.

Vessels with a conical or dished bottom require a somewhat more complicated relationship.

For vessels with internals such as cooling or heating coils, the vessel fabricator provides a graph or table (often called the strapping table), the options being as follows:

1. Relate volume to vessel level. The volume is multiplied by fluid density to obtain mass of the contents.
2. Relate mass of water (at 4°C) to vessel level. The mass of water is multiplied by fluid-specific gravity to obtain mass of the contents.

Either relationship is easily entered into digital control systems using a characterization function or function generator block.

Occasionally, level is the measure of interest. For example, when charging a heel to a vessel, a sufficient quantity must be added to adequately submerge the agitator. Whether the agitator is submerged or not depends on the level of liquid within the vessel. However, high precision is not generally required.

Level applies to solid surfaces as well as liquid surfaces. But for solids, issues such as the following complicate relating weight to level:

1. The surface may not be even. When adding solids, a distributor will provide a more even surface. Otherwise, solids tend to "mound up." When removing solids from the bottom of the vessel, solids tend to be preferentially removed from the center, resulting in an inverted cone at the surface.
2. Stratification can be a problem in liquids, but even mild agitation (such as a pump-around loop) will avoid the worst consequences of stratification. But for solids, mixing is not realistic, making stratification inevitable.
3. Air spaces known as "rat holes" can form beneath the surface of the solids. These increase the level of the solids in the vessel, but not the weight of the solids.

For solids, weight is usually far preferable to level.

But there is an exception. Where the concern is to avoid overfilling or completely emptying, the focus must be on level. Fortunately, level switches are normally adequate. Products based on vibrating elements, ultrasonics, and capacitance are applicable to both liquids and solids. But especially for solids, the preferable technology depends on the nature of the solids.

2.3.2. Load Cells

A load cell consists of two components:

1. A structural element for which the deformation (change in physical size) can be accurately related to the force being applied to the structural element.
2. Strain gauges attached to the structural element convert the deformation into an electrical property that can be sensed.

Two configurations for incorporating one or more load cells into the vessel supports are possible:

Tension. Solid materials are often discharged into a pan suspended from a single support via a load cell. The strain gauges of the load cell are in tension. Operating experience with this type of load cell has generally been excellent.

Compression. Vessels are normally mounted on three or four supports. A load cell is incorporated into each support. The strain gauges of the load cells are in compression. The load cells indicate the weight on each support; from these, the weight transmitter computes the measured value of the vessel weight. When installed on feed tanks, reactors, and the like, compromises seem to be inevitable, causing the load cell performance to suffer.

Load cells cannot be installed on tanks whose bottom rests on the ground (such as the large tanks in the tank farms used in oil refining), for buried tanks (such as those at service stations), and so on. Fortunately, these are not common in batch facilities.

Under ideal situations, the accuracy of the weight measurement based on load cells is excellent. But process applications often depart from ideal. The more difficult the application, the greater the departure. The following are major factors that impair the accuracy of weight measurements in process applications:

1. The load cells sense the total weight of the vessel and its contents. For heavy vessels, the load cell must sense a small change in a large value. The weight transmitter can also include a bias or "tare" so that the measured value is the net weight of the contents. But do not be deceived; the load cells are sensing total vessel weight. For small changes in weight,

the net result is to subtract two large numbers to obtain a small one. Such numerical calculations amplify errors in either large number.

2. Load cells can be installed on jacketed vessels, but any change in the weight of the fluid in the jacket is interpreted as a change in the weight of the contents of the vessel.

3. Avoiding stresses imposed by piping and other structural requirements usually proves difficult. Load cells work best when the force on the load cell is entirely vertical. As horizontal stresses are introduced, the load cells are affected. These issues are easiest to address when the load cell is incorporated into the original vessel design and installation.

How to properly incorporate load cells into the vessel supports is well known. But especially when retrofitting load cells to existing vessels, the result is not always what one would like. For example, how do you conform to the recommended installation procedures when the available vertical height is inadequate?

Although flexible connections may be recommended, objections are likely to arise for hazardous materials (even relatively benign materials but at high pressure or temperature). Personnel safety always trumps recommended practices for installing load cells. In making compromises, there is no substitute for experience. For example, adding metallic braiding to improve the strength of a flexible connection can result in a connection that is flexible in directions perpendicular to the piping, but not in the piping direction.

The consequences of rigid piping on load cells include the following:

1. Horizontal stresses change with temperature due to the expansion and contraction of metals.

2. Vibrations from equipment such as pumps and agitators are propagated to the vessel.

These degrade the performance of the load cells, but how much is difficult to quantify. The normal approach for calibrating the load cells is to attach known weights to the vessel supports and verify that the load cells respond accurately. Such procedures do not quantify the impact of compromises such as rigid piping.

2.3.3. Noise

Weight measurements from load cells are invariably accompanied by noise. Two components are usually present:

1. Noise originating within the load cells and associated electronics. As this noise is usually high frequency in nature, smoothing must be incorporated into the weight transmitter to address this noise. Most expect the

manufacturer of the weight measurement to address this component of the noise.

2. Noise entering through the vessel supports. The characteristics of this noise vary from one installation to the next. Rarely is such noise entirely random, but often exhibits cycles that are evident in trend plots. Vessels mounted on supports vibrate at a natural harmonic whose frequency decreases with the size of the vessel. The rotational speed of agitators in nearby vessels is another source of a low-frequency component of the noise.

Characterizing the noise in a weight measurement requires a spectral analysis. The signal for the weight measurement is expressed as the sum of the frequencies of its individual components. The results are usually presented as a graph of the amplitude as a function of frequency. While this technology is well known, frequency analysis is not popular within the process industries.

As compared with other industries, the data rates within the process industries are relatively slow. Typical scan rates are 1 or 2 samples/second (scan interval of 1 second or 0.5 seconds). Such scan rates are inadequate for a spectral analysis of a weight measurement. Scan rates of 10 samples/second are used in a few applications, but even this is only marginally adequate for a spectral analysis.

The data rate issues extend beyond the scan rate of the data acquisition equipment. Being microprocessor-based, the weight transmitter updates the current value of the weight measurement on some interval. Scanning faster that the weight measurement is updated makes no sense.

Especially as compared with communications and related industries, the approach within the process industries to filtering and smoothing is very rudimentary. The typical practice is summarized as follows:

If the measured value is bouncing too much, increase the smoothing.

Most filters are of the exponential or lag variety. Occasionally, a moving average filter is used, but these perform much like the exponential filter. Rarely is any other type of filter applied. Implementing a high-performance filter is only part of the issue. To get the desired results from such a filter, its coefficients must be set properly, which requires a spectral analysis of the noise.

2.3.4. Lag Filters

Some smoothing is required when converting the signals from the load cells into a weight measurement. Most weight transmitters permit the user to specify additional smoothing.

Additional smoothing can also be applied within the input processing for the controls. This smoothing is usually in the form of the customary exponential or lag filter that is described by the following first-order differential equation:

$$\tau_F \frac{dy(t)}{dt} + y(t) = x(t),$$

where

$y(t)$ = Smoothed value at time t

$x(t)$ = Raw value at time t

t = Time (second or minute)

τ_F = Filter time constant (second or minute).

This equation is normally implemented as a difference equation with a smoothing coefficient k:

$$y_i = kx_i + (1-k)y_{i-1} = y_{i-1} + k(x_i - y_{i-1})$$

$$k = 1 - \exp\left(-\frac{\Delta t}{\tau_F}\right) \cong \frac{\Delta t}{\tau_F + \Delta t}.$$

That the term "lag filter" is appropriate should be evident from Figure 2.1. The response is to a ramp change in the input to the filter. Initially, the weight of the vessel is 800 kg. At time 0, a discharge flow of 40 kg/min commences, which causes the value of the weight to change in a ramp fashion with a slope of −40 kg/min. After 15 minutes, the discharge flow is stopped, which gives a final vessel weight of 200 kg.

The value of the vessel weight is the input to a filter with a time constant of 1.0 minute. After approximately five filter time constants (5 minutes for

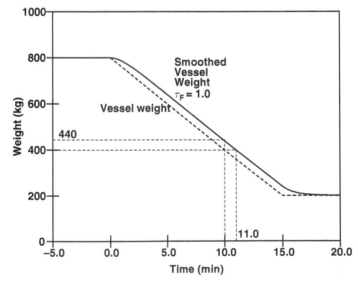

Figure 2.1. Response of the exponential or lag filter to a ramp change in its input.

Figure 2.1), the rate of change of the smoothed value is -40 kg/min, which is the same as the rate of change of the vessel weight. However, the smoothed value clearly lags behind the vessel weight. After 10 minutes, the weight of the vessel has changed from 800 kg to 400 kg. However, the smoothed value exhibits a "lag" in two respects:

1. The smoothed value attains 400 kg in 11 minutes, which is 1 minute later. The output of the filter is lagging behind the vessel weight by a time equal to the filter time constant (1.0 minute for Figure 2.1).
2. After 10 minutes, the smoothed value is 440 kg, which is greater than the vessel weight. The output of the filter is lagging behind the vessel weight by an amount equal to the product of the following:
   (a) The rate of change of the vessel weight (-40 kg/min for Figure 2.1).
   (b) The filter time constant (1.0 minute for Figure 2.1).

This has implications for material transfers.

2.3.5. Material Transfers

If the weight of the vessel is constant, the smoothed value will approach the vessel weight. For Figure 2.1, the input flow is stopped after 15 minutes, which gives a vessel weight of 200 kg. But at $t = 15$ minutes, the smoothed vessel weight is 240 kg. After approximately five filter time constants (5 minutes for Figure 2.1), the smoothed vessel weight is essentially the same as the vessel weight.

The amount of material transferred into the vessel can be computed from the vessel weight prior to the start of the transfer and the vessel weight after the transfer is terminated. When noise is present on the measured value of the weight, smoothed values must be used. The most common options are the following:

1. Use the smoothed value from the exponential or lag filter. One must wait until the smoothed value "ceases to change," which means approximately five filter time constants following the termination of a material transfer.
2. Compute the arithmetic average of the weight input over some time interval (1.0 minute seems to be a common choice). Typical approaches include the following:
   (a) Wait one time interval after the termination of a material transfer and then compute the arithmetic average over the next time interval.
   (b) Compare the computed value of the arithmetic average with the value from the previous time interval. Continue computing arithmetic averages until two consecutive values differ by less than a specified tolerance.

These considerations apply to the initial value of the vessel weight as well as the final value of the vessel weight.

For batch operations, an important consideration is the time required to obtain the values of the initial and final vessel weights. In most (but not all) cases, the additional time required to obtain the initial and final values increases the time required to manufacture a batch, which translates to a decrease in the production rate for the plant. At least for certain material transfers, improving the accuracy decreases the batch-to-batch variability of product quality. However, such benefits come at the cost of a decreased production rate.

2.3.6. Noise on Vessel Weight Measurement

Figure 2.2 illustrates the application of exponential smoothing to a weight measurement that is accompanied by noise. The material transfer is out of the vessel. The initial weight is 800 kg; material flows out at a rate of 40 kg/min; the final weight is 200 kg. Using a filter time constant of 1.0 minute removes most of the variability from the smoothed vessel weight.

However, the lag in the filter output is the same as in Figure 2.1. Due to the lag contributed by the filter, the smoothed vessel weight is approximately 40 kg greater than the actual vessel weight.

Prior to and subsequent to the material transfer, the vessel weight is constant, which permits an arithmetic average to be used to obtain a smoothed value. But during the material transfer, does using an arithmetic average make sense? At least intuitively, the answer is "no."

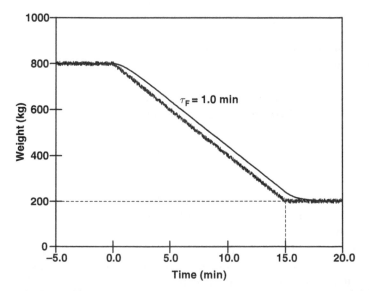

Figure 2.2. Smoothing a noisy weight measurement using an exponential filter.

2.3.7. Moving Average Filter

An alternative to the exponential filter is the moving average filter, which computes the arithmetic average of the input values over an interval that can be specified as either the time duration or the number of samples (60 readings at a 1-second interval gives a time interval of 1.0 minute). The term "moving" means that the average is recomputed every time a new reading becomes available (if the interval between readings is 1 second, the average is recomputed on a 1-second interval). This requires a storage array for the number of previous values required to compute the average. Early digital control products provided limited memory, so reserving the necessary storage for the moving average filter was a problem. The exponential filter requires no such storage, so it quickly became the filter of choice and was provided by virtually all digital control products. But with modern controls, memory limitations have become a nonissue, and most suppliers now provide a moving average filter.

In practice, the differences between these two are not that great, as illustrated by the following two figures:

Figure 2.2. Smoothing is provided by an exponential filter with a time constant τ_F of 1.0 minute.

Figure 2.3. Smoothing is provided by a moving average filter with an averaging time T_A of 2.0 minute.

The smoothed values in these two figures are not identical, but the differences are minor. Provided $T_A = 2\,\tau_F$, the behavior of the moving average filter and the exponential filter are essentially the same.

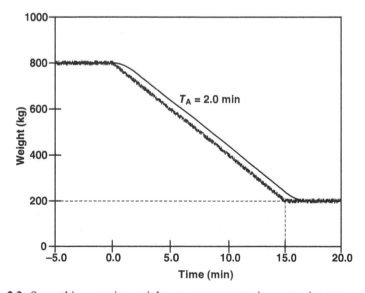

Figure 2.3. Smoothing a noisy weight measurement using a moving average filter.

In a sense, the exponential filter is also an averaging filter. The arithmetic average weighs all points equally over the interval for which the average is computed. The exponential filter uses an exponential weighting function, with recent points weighed more heavily than older points. This aspect seems reasonable, but another aspect does not. Even very old points contribute to the average, albeit in a very small way. Theoretically, the weighting value for a point never decreases to zero, but in practice, points older than five filter time constants in the past contribute very little to the average.

Two observations have been made:

1. While a material transfer is in progress, computing an arithmetic average of the points over a specified interval (the moving average filter) does not intuitively seem appropriate.
2. For $T_A = 2\,\tau_F$, the performance of the exponential filter is essentially the same as the moving average filter (Figure 2.2 and Figure 2.3).

These observations suggest that applying the exponential filter during a material transfer is at least questionable. By weighting recent values more heavily than older values, the exponential filter may be preferable to a moving average filter that weighs all values equally. But in the end, both are averaging filters.

2.3.8. Vessel Weight during a Material Transfer

Consider the following logic for terminating a material transfer:

1. Record the initial weight of the vessel.
2. To obtain a target for the vessel weight at which to terminate the material transfer, add to or subtract from the vessel weight the amount of the material to be transferred.
3. Start the material transfer.
4. When the vessel weight attains the target, terminate the material transfer.

Suppose the smoothed value for the vessel weight is used in this logic. Figure 2.4 illustrates terminating the material transfer based on the smoothed value from an exponential filter with a filter time constant of 0.5 minute. The smoothed value attains the target of 200 kg at approximately 15.5 minutes. The smoothed value is approximately 20 kg higher than the actual vessel weight (20 kg = 40 kg/min × 0.5 minute). The material transfer should be terminated at $t = 15$ minutes, but based on the smoothed value, the transfer is terminated at approximately 15.5 minutes. Consequently, the final vessel weight is approximately 180 kg, which is below the target by 20 kg.

When material transfers are terminated based on a smoothed value, the smallest degree of smoothing should be used. Figure 2.4 also illustrates

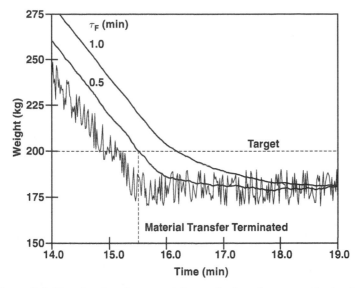

Figure 2.4. Terminating the material transfer based on smoothed value.

smoothing using a time constant of 1.0 minute. The smoothed value exhibits less variability, but the smoothed value is approximately 40 kg higher than the actual vessel weight. Terminating the material transfer based on this smoothed value would give a final value of approximately 160 kg.

Instead of terminating the transfer when the smoothed value attains the target, the transfer should be terminated earlier. The target for terminating the transfer is computed as discussed previously, but then a small amount is added or subtracted to give the value (sometimes called the preset) at which the material transfer is terminated. How much to add or subtract depends on several factors:

1. Smoothing, as illustrated in Figure 2.4.
2. Valve travel time. Small valves with electric or pneumatic actuators move very quickly, but there are exceptions.
3. Fluid drained or purged from piping. This should be a small amount, but as always, there are exceptions.

The value to add or subtract must be adjusted based on the errors resulting from previous material transfers. If the same material is transferred at the same rate, a good value can usually be determined.

As will be discussed in a subsequent chapter, the accuracy of a material transfer can be improved by using a "dribble flow." Most of the material is transferred as rapidly as possible, but upon approaching the target, the flow is reduced to a small value.

2.3.9. Least Squares Filter

This filter functions as follows:

1. The smoothed value is computed from the current input value and some number of previous input values. This requires a storage array in the same manner as the moving average filter. Let n be the number of values used to compute the smoothed value.
2. A linear least squares computation is used to fit a straight line to the data, the results being the slope and intercept for the line.

In early digital controls, both the storage for the n data values and the additional computational burden presented problems. However, modern digital controls are easily up to the task.

The preferable way to formulate the linear least squares computation is to designate time 0 as time of the most recent point. The data value will be designated as Y_0. The time for the previous point would be $-\Delta t$, where Δt is the interval between data points. The data value will be designated as Y_1. The X and Y values for the linear least squares fit are as follows:

Index	X	Y
0	0	Y_0
1	$-\Delta t$	Y_1
2	$-2\Delta t$	Y_2
$n-1$	$-(n-1)\Delta t$	Y_{n-1}

When formulated in this manner, the coefficients have the following significance:

Intercept. Smoothed value for the weight.

Slope. Smoothed value for the flow. A positive value for the slope means that material is flowing into the vessel; a negative value for the slope means that material is flowing out of the vessel.

Attention will initially be restricted to the smoothed value of the weight, but the smoothed value of the flow will be examined shortly.

The least squares computation requires computing the following sums:

$$\text{Sum of } X: \quad S_X = \sum_{j=0}^{n-1} X_j$$

$$\text{Sum of } X^2: \quad S_{XX} = \sum_{j=0}^{n-1} X_j^2$$

$$\text{Sum of } Y: \qquad S_Y = \sum_{j=0}^{n-1} Y_j$$

$$\text{Sum of } X \times Y: \quad S_{XY} = \sum_{j=0}^{n-1} X_j Y_j.$$

The values for S_X and S_{XX} depend only on the values of n and Δt, so values for these can be computed once and saved. However, values for S_Y and S_{XY} must be recomputed on each execution of the least squares filter.

From the values of these sums, the intercept b and slope a are computed as follows:

$$a = \frac{n\, S_{XY} - S_X\, S_Y}{n\, S_{XX} - S_X{}^2}$$

$$b = \frac{S_Y - a\, S_X}{n}.$$

The arithmetic moving average (the output of a moving average filter) is easily computed as S_Y/n. This creates the following possibilities:

Transfer in progress: Use b for the smoothed value of the weight.

Transfer not in progress: Use the arithmetic average for the weight.

Figure 2.5 illustrates the performance of the least squares filter for two cases:

$n = 60$. For $\Delta t = 1$ second, the time interval for the smoothing is 1.0 minute.

$n = 120$. For $\Delta t = 1$ second, the time interval for the smoothing is 2.0 minutes (same as for the moving average filter in Figure 2.3).

The following points should be noted:

1. As compared with the exponential filter and the moving average filter, the least squares filter does not exhibit the lag during a material transfer. Terminating the material transfer based on the smoothed value from the least squares filter occurs at approximately the correct time ($t = 15$ minutes in Figure 2.5). A preset is not required to compensate for the lag from the least squares filter, but could still be required for other smoothing, for valve travel time, for material draining from the transfer piping, and so on.
2. Following the termination of the material transfer, at least one smoothing time is required for the smoothed value to approach the final vessel weight. This is especially evident from Figure 2.5b. Following the

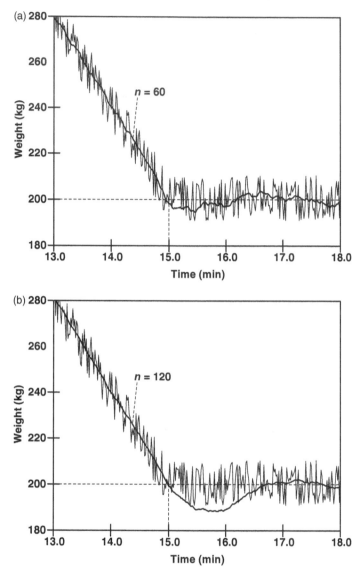

Figure 2.5. Smoothing a noisy weight measurement using a least squares filter: (a) smoothing time interval of 1.0 minute (60 points); (b) smoothing time interval of 2.0 minutes (120 points).

termination of the material transfer, the smoothed value of the weight drops significantly below the final weight. But after one smoothing time (1.0 minute for Figure 2.5a or 2.0 minutes for Figure 2.5b), the smoothed value of the weight is close to the final vessel weight. A similar phenomenon occurs at the beginning of the material transfer—one smoothing time is required for the smoothed value from the least squares filter to approach the actual vessel weight.

Another potential application of the least squares filter is to compute the rate of change for critical variables such as reactor temperatures. Temperatures from probes inserted into thermowells are accompanied by very little noise, but computing the rate of change from unsmoothed temperature values tends to give erratic values for the rate of change. Rates of change computed from smoothed values from an exponential filter introduce the same lag as described earlier.

2.4. FLOW MEASUREMENTS

Within batch facilities, flow measurements are taken very seriously. The reaction stoichiometry ultimately determines the values for the various flows. The possibilities are illustrated by the following two cases:

1. A given reactant is fed in excess. Subsequent separation equipment recovers the excess charge, which is recycled and used as feed to subsequent batches. The consequences of small errors in the feed of this reactant are usually nominal.
2. A given reactant is fed in a stoichiometric ratio to another reactant. When no subsequent separation (or purification) occurs, any error in either feed results in unreacted material remaining in the reactor products, usually with implications on the product quality.

Especially for the latter, the amounts charged must be accurate.

2.4.1. Mass Flow

Reaction stoichiometry is based on molar quantities, which are easily and precisely converted to mass quantities. Consequently, measurement devices that directly sense mass or weight are preferable. In industrial plants, only two options are practical:

Coriolis flow meters. These are the only mass meters routinely installed in industrial facilities. The flow is measured in kilogram/minute, pound/minute, and so on.

Load cells. These sense the total weight of a vessel and its contents. The flow is the rate of change of vessel weight, which is the basis for the loss-in-weight feeders that will be discussed in the next section of this chapter.

Magnetic flow meters, vortex shedding flow meters, ultrasonic flow meters, head-type flow meters, rotameters, and so on, are volumetric meters that directly sense the volumetric flow as liters/minute, gallon/minute, and so on. Multiplying the volumetric flow by the fluid density gives the mass flow; however, errors in the fluid density introduce errors into the computed value for the mass flow. Changes in fluid temperature and/or composition affect the fluid density.

2.4.2. Coriolis Meters

A company named Micromotion pioneered this technology, and these meters are sometimes referred to as Micromotion meters. The principle on which these meters rely is the coriolis principle, so herein they will be referred to as coriolis meters.

Coriolis meters are now available from a variety of suppliers and in a variety of tube configurations, including a straight tube. These meters can be applied to the following:

1. Gas flow
2. Liquid flow
3. Two-phase flow, provided neither phase is compressible. Specifically, the meters can sense the mass flow of slurries (a solid and liquid phase) and emulsions (two liquid phases).

Coriolis meters are the meter of choice in most applications in batch facilities. The most obvious advantage is that the meter senses mass flow. Its accuracy rivals that of load cells as they are installed in production facilities. The measured mass flow is not affected by fluid properties (density, viscosity, etc.), which is also a major advantage where a single meter is used to measure the flow of different materials. In batch applications, the flow of a given material may be large for one product but small for another. The coriolis meter's turndown ratio is in excess of 10:1, which is a distinct advantage in such applications.

The coriolis meter can sense some flows that were previously unmeasurable, one example being the flow of an emulsion. The emulsion is prepared in a feed tank and then transferred to the main reactor at a proscribed rate, with the flow of one or more co-feeds being a specified ratio to the emulsion flow. Prior to the introduction of the coriolis meter, the flow meter was basically a process operator with a sight glass and a stopwatch. At the start of the emulsion feed, the level in the sight glass should be a specified value. After 30 minutes, the level in the sight glass should be another specified value. Although the operators proved rather adept at getting the desired emulsion flow (especially according to the data they recorded), this approach has obvious limitations. Coriolis meters can measure the mass flow of the emulsion stream, improving the accuracy of the emulsion feed to the reactor as well as opening the possibility of controlling the reactor temperature by adjusting the emulsion flow (along with the co-feeds).

The major disadvantage of the coriolis meter is its cost. But costs have declined as other manufacturers introduced products. It tends to be a bulky meter (especially in large sizes) as well as requiring a large pressure drop. The straight-tube versions of the coriolis meter address these problems, but with some loss of accuracy. To date, straight-tube meters are not common in batch facilities, where accuracy is very important, and the issues associated with the bulky nature of the meter and the high pressure drop can be tolerated.

Probably the main problem with the meter in liquid flow applications is the presence of gas bubbles within the flow stream. Upstream agitators or pumps can impart some of this to a flowing liquid. As the pressure reduces, dissolved gases can come out of solution. Even a small amount of gas in a liquid stream adversely affects a coriolis meter. The installation of the coriolis meter must be such that no gas pockets can be trapped within the meter.

If a coriolis meter is not accurately measuring the flow, the first issue raised by the manufacturer is the presence of gas bubbles or pockets within the meter. In designing the piping for the coriolis meter, assume that this question will be raised and be prepared to answer it. Factors such as mechanical stress on the meter assembly can also affect its accuracy, but the issue of gas bubbles or pockets always arises first.

2.4.3. Density

The coriolis meter can sense the fluid density as well as the mass flow. Dividing the mass flow by the fluid density gives the volumetric flow, but this is rarely of interest in a batch facility (except possibly in conjunction with a check meter, as will be discussed in a subsequent chapter). However, other possibilities arise in batch facilities.

To feed a specified amount of a material to a vessel, the usual procedure is to open the appropriate valves, start the transfer pump, continue until the desired quantity of the material has been transferred, stop the pump, and close the valves. Depending on the piping arrangement, stopping the pump permits the liquid within the piping to "bleed back," leaving air within the piping and the flow meter. At the start of the next transfer, air flows through the meter until all is displaced by liquid. The mass flow of the air is not to be included in the flow total. Furthermore, a coriolis meter designed for liquids does not necessarily give a zero flow reading when filled with air. Both of these are addressed with a "low-density cutoff." That is, if the fluid density sensed by the coriolis meter is less than a specified value, the flow indication is zero. This logic is normally implemented within the transmitter (all are microprocessor based) for the coriolis meter.

In batch facilities the separation of an aqueous phase from an organic phase is commonly accomplished by a batch decanter. The mixture is transferred to the decanter, allowed to separate for the proscribed time, and then the dense phase drained from the bottom of the decanter. By draining through a coriolis meter, the density of the material flowing through the meter can be sensed, and the flow stopped when the density decreases (indicating the less dense phase). Other possibilities exist for sensing the density, but with coriolis meters widely installed for measuring flows, using the same equipment to measure density has a definite appeal.

A major concern in batch facilities can be stated as follows: Is the material being charged actually the material we think it is? If not, the obvious consequence is that the desired product will not be realized. However, far more serious consequences could arise. Will there be a reaction, and if so, what reaction? If the latter, the equipment, including pressure relief systems, is not designed for this reaction. The possibility exists for numerous consequences, most of which are bad. These situations do not arise very often, but when they do, they generally break the monotony of the day.

By sensing fluid density, the coriolis meter offers the possibility of detecting some of these situations and aborting the material transfer. A narrow range can be specified for the density of the material that should be flowing. If the sensed density is outside this range, the material transfer is stopped immediately. Clearly this is not foolproof, but the necessary logic is easily incorporated into the controls and increases the confidence that we are actually doing what we intend to be doing.

2.4.4. Heating or Cooling Media Flows

A heating media such as hot oil or a cooling media such as cooling water adds or removes heat in the form of the sensible heat of the fluid. Most such fluids are relatively benign. The flow measurement is performed at pressures and temperatures that are well within the limits of the metering technology being applied. The accuracy requirements are much lower than for the flows of materials charged to the process. For this application, volumetric meters are acceptable.

At least at first glance, this seems to be a straightforward metering application.

But in batch facilities, a complication arises. Especially for units such as reactors, the heat transfer rate varies considerably throughout the batch. Usually, the maximum heat transfer rate occurs in the early stages of the batch. Where co-feeds are involved, the rate of reaction and the associated heat transfer rate depend largely on the feed rate for the co-feeds. But after all feeds are stopped, the reaction rate and consequently the heat transfer rate decrease rapidly. Especially when the reaction must be driven to completion, the heat transfer rate near the end of the batch is much lower than during the early stages of the batch.

A flow meter must be sized for the largest flow rate that must be metered. But for metering the heating or cooling media, the turndown ratio of the flow meter must not be ignored. For a flow meter, the turndown ratio is the ratio of the largest flow that can be accurately measured (usually the upper range value for the meter) to the smallest nonzero flow that can be accurately measured. For heating or cooling media flow meters, a value must be obtained for the smallest flow rate that must be measurable. The required turndown ratio is the ratio of the largest flow rate to the smallest flow rate.

For such applications, head-type flow meters such as the orifice meter are rarely acceptable. Because the pressure drop is proportional to the square of the flow, these meters have a turndown ratio of about 4:1. Even installing "smart" differential pressure transmitters does not significantly increase the turndown ratio—the limitation is imposed by the metering principle. At low flow rates, the differential pressure is small, giving a low signal-to-noise ratio. The linearization provided by a square-root extractor amplifies the noise, giving an erratic measurement at low flows.

To obtain a high turndown ratio, meters with linear characteristics are required. Magnetic flow meters, ultrasonic meters, and vortex-shedding meters are linear meters with turndown ratios of 10:1 or higher. However, a vortex-shedding meter has another requirement that must not be ignored. Turndown ratios stated for vortex-shedding meters tend to be high, but the "fine-print" states that the flow regime must always be turbulent. If the flow regime is laminar, the output of a vortex-shedding meter is zero. Manufacturers of vortex meters clearly state the maximum flow that can be measured and the minimum flow that can be measured by a given meter. The vortex-shedding meter only works between these two values, and both must be taken seriously. As the heating or cooling media flow is reduced from a large value, the output of the flow transmitter tracks the flow very well until the flow regime changes to laminar, at which time the transmitter output abruptly drops to zero. Meters such as the magnetic flow meter may be inaccurate at low flow rates, but the measured value for the flow is not zero.

2.4.5. Coriolis Meters versus Load Cells

For chemicals and similar materials that are transported by tank trucks, the routine practice is to weigh the truck on entry to the plant and departure from the plant, the difference in weight being the amount of material shipped or received. The weight is determined by what is essentially a very large platform scale that relies on load cell technology to determine the weight of the vehicle. Whenever payments are based on a measurement, the most accurate measurement technology available is used.

For weight measurements, the load cell is very accurate, but only under ideal conditions. Truck scales meet these requirements. But as discussed in a previous section of this chapter, process installations of load cells usually fall short, especially for process vessels that are mounted on load cells. For liquids flowing into or out of such vessels, coriolis meters are often installed. This gives two options for determining the total amount of material transferred:

1. Difference in the weight measurement before and after the material transfer.

2. Flow totalizer whose input is the flow measurement from the coriolis meter.

One is tempted to simply assume that the difference in the weight measurements is the most accurate value for the total amount of material transferred, but ultimately this depends on the compromises made in the load cell installation.

The degree of agreement (or disagreement) between the two measurements can be assessed by procedures such as the following:

1. Stop all agitators, pumps, and so on. Then wait until the weight measurement ceases to change.
2. Restart agitators, pumps, and so on.
3. Execute a material transfer through the coriolis meter, terminating based on the totalized value for the flow.
4. Stop all agitators, pumps, and so on. Then wait until the weight measurement ceases to change.

Sometimes this is possible for certain material transfers that are executed for a production batch. However, stopping agitators, cooling water pumps, and so on, is not permitted when reactive materials are in a vessel. In such cases, the tests must be performed between batches using materials such as water.

The difference in the weight measurements usually agrees quite closely with the total flow computed from the coriolis meter, so either could be used. However, the coriolis meter has a major advantage—there is no need to delay production operations before and after the material transfer. Doing so for every material transfer would noticeably increase the time to manufacture a batch of product.

In some cases, a coriolis meter is installed on a material transfer that was previously executed based on weight measurements. Before switching to using the same flow total for the coriolis meter, execute each material transfer based on the weight measurements and note the flow total from the coriolis meter. In most cases, the flow total based on the coriolis meter is closer to the actual amount of material that was transferred. The ratio of the two amounts is often defined as a meter factor and is used to scale all transfers. Unfortunately, the value of the meter factor could depend on the material being transferred.

When the two values from the test are significantly different (meter factor is significantly different from 1.0), one is faced with two options:

1. Continue using the same target for the material transfer amount. The likely impact is a change in product properties, which will surely come to the attention of the statistical quality control (SQC) people.
2. Use the amount that the coriolis meter indicated as being transferred during the test as the target for future material transfers. This will

minimize the impact on the product properties, but the change in the target amount will require some explanation. This can have political as well as technical overtones.

2.5. LOSS-IN-WEIGHT APPLICATION

The objective of a "loss-in-weight" application is to discharge material from a vessel at a specified rate based on a measurement of the vessel weight. "Loss in weight" implies that material is being removed from the vessel, which is the more common case. However, the same issues apply when material is being transferred into a vessel.

The options for controlling the flow of the material being discharged from a vessel are as follows:

Liquid. Loss in weight is applicable, but where practical, installing a coriolis meter to measure a liquid flow is usually a preferred alternative.

Solid. Metering a solid flow is usually impractical, making loss in weight the only viable option.

Often the output from the loss-in-weight feeder is the speed of a rotary valve, a screw feeder, or other mechanism that determines the rate of discharge of the solids material.

2.5.1. Weight to Flow

The simplest approach for converting from weight to flow is to use finite differences:

$$F_i \cong \frac{W_i - W_{i-1}}{\Delta t},$$

where

F_i = Average flow over time interval i
W_i = Weight at end of time interval i
W_{i-1} = Weight at end of time interval $i-1$ (or beginning of time interval i)
Δt = Duration of time interval.

Finite difference computations involve subtracting two large numbers to obtain a small number, which raises serious numerical issues especially when applied to real-time data.

Over long time intervals (like 10 minutes), the finite difference computation can convert the weight readings to the average flow without excessively amplifying the noise. But what if the objective is to compute the instantaneous flow? For process applications, the average flow over a one-second interval is usually acceptable as the "instantaneous flow." Applying the finite difference calculation over a 1-second interval is not practical—any noise in the weight measurement is greatly amplified in the value computed for the flow.

2.5.2. Exponential Smoothing

This can be applied in the following ways:

1. Smooth the weight measurement and then apply finite differences to compute the flow.
2. Apply finite differences to compute the flow from the weight measurement and then smooth the values computed for the flow.

The first approach is most common.

Figure 2.2 presents the smoothed weight obtained using an exponential filter with a filter time constant of 1.0 minute. Figure 2.6a presents the flow computed from the smoothed weight using finite differences with a time interval of 1.0 second. Very little variability is evident in the smoothed weight in Figure 2.2. However, the finite difference calculation greatly amplifies this variability, giving the very noisy flow presented in Figure 2.6a.

Increasing the time interval used to compute the finite differences decreases the amplification of the noise. Figure 2.6b presents the flow computed using a time interval of 60 seconds for computing the finite difference. The noise in the flow is much less, but there is a downside.

Increasing the smoothing and increasing the time interval for computing the finite difference have one aspect in common—both introduce lag in the computed value for the flow. The actual discharge flow increases abruptly from 0.0 to 40.0 kg/min at $t = 0$, and then decreases abruptly from 40.0 to 0.0 kg/min at $t = 15$. Despite the noise, a lag is evident in the flow in Figure 2.6a. This lag is due to the exponential smoothing applied to the weight measurement. Figure 2.6b exhibits an additional lag of approximately 30 seconds that is due to computing the flow using a time interval of 60 seconds for the finite difference calculation.

2.5.3. Least Squares Filter

The least squares filter computes the slope and intercept of a straight-line fit to the weight measurements over a specified interval (the smoothing or averaging time). As previously formulated, the slope computed by the least squares filter is the net flow into the vessel.

Figure 2.5b illustrated the smoothed weight computed using a least squares filter with $n = 120$ (time interval of 2.0 minutes). Figure 2.5b plots the values computed for the intercept; Figure 2.7 plots the values computed for the slope

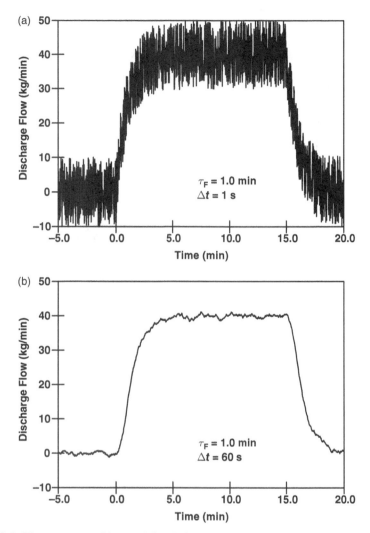

Figure 2.6. Flow computed by applying finite differences to the smoothed weight from Figure 2.2: (a) time interval of 1 second; (b) time interval of 60 seconds.

(or actually the negative of the slope to obtain the discharge flow). Starting at $t = 0$, the flow changes in essentially a ramp fashion from 0.0 kg/min to 40.0 kg/min over a time interval of 2.0 minutes. At $t = 15$, the behavior is similar, but in the opposite direction.

The least squares filter also exhibits a lag in the sense that the value computed for the flow does not change abruptly. Following any major change in the discharge flow, the computed values will be in error for one smoothing time, but thereafter the values computed for the discharge flow will be close to the actual discharge flow.

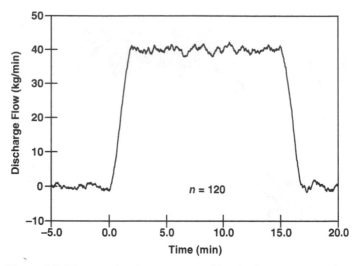

Figure 2.7. Flow as the slope computed by the least squares filter.

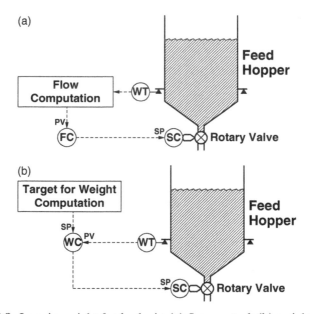

Figure 2.8. Loss-in-weight feeder logic: (a) flow control; (b) weight control.

2.5.4. Control Alternatives

One possible configuration for the control logic is the flow control configuration illustrated in Figure 2.8a. The value of the flow must be derived from the weight measurement. The difficulty of doing this for a noisy weight measurement was discussed earlier.

An alternate configuration for the control logic is the weight control configuration illustrated in Figure 2.8b. The target for the vessel weight is decreased in a ramp fashion, the slope of the ramp being the desired discharge weight. Although it is not necessary to derive the discharge flow from the weight measurement, any smoothing of the weight measurement introduces lag into the smoothed value.

The weight controller is essentially a level controller. The process is integrating—the discharge flow is the rate of change of the vessel weight. Including the reset mode in the controller means the loop contains two integrators—one in the controller and one in the process. This raises the same tuning issues as for level controllers:

1. If the reset time is shorter than the total lag in the process, the loop is unstable for all values of the controller gain. Fortunately, lags in loss-in-weight feeders tend to be short. However, the lag introduced through smoothing the weight measurement must also be included in the total lag.
2. If the degree of cycling in a loop is excessive, the usual action is to reduce the controller gain. But for loops with double integrators and short reset times, reducing the controller gain has two undesirable consequences:
   (a) The cycle decays more slowly.
   (b) The period of the cycle increases.

A pure integrating process coupled with an integral-only controller results in a loop that cycles continuously. When a proportional-integral (PI) controller is applied to an integrating process, some proportional action is required to obtain a stable loop.

For loss-in-weight feeder applications, there is one advantage to the process being integrating. A PI controller is able to track a ramp change in the set point with no offset or lag. For a nonintegrating process, a PI controller cannot track a ramp change in the controller set point without offset or lag.

The noisy weight measurement coupled with the integrating nature of the process presents a dilemma in tuning. For a noisy process variable input, the options are as follows:

1. Smooth the input and then apply the PI control logic to the smoothed value. Unfortunately, the lag introduced by the smoothing will require that the controller be tuned more conservatively.
2. Minimize the smoothing of the input and rely primarily on the reset mode for control action. The controller gain must be low so as to not amplify the noise. In a sense, the reset mode is providing both smoothing and control action.

When the process is integrating (as in a loss-in-weight feeder), the latter approach conflicts with the tuning considerations. For integrating processes (the most common being level loops), the recommended approach is to use a long reset time and adjust the controller gain to give the desired performance.

REFERENCES

1. Smith, C. L., *Basic Process Measurements*, John Wiley and Sons, New York, 2009.
2. Smith, C. L., *Advanced Process Control: Beyond Single Loop Control*, John Wiley and Sons, New York, 2010.

3

CONTINUOUS CONTROL ISSUES

The focus of this chapter is on continuous control issues that are especially important in batch applications. Although they can arise in continuous processes, the following three issues arise in batch facilities either more frequently or to a greater degree:

Loops that operate intermittently. For most loops in a batch facility, the proportional-integral-derivative (PID) control calculations are providing closed loop control only during certain periods during the batch. At other times, the loop is disabled, either by switching the controller to manual, by activating tracking, by closing a block valve, or other. When closed loop control resumes, a smooth or bumpless transition is essential.

Widely varying process conditions. Heat transfer rates in batch reactors can vary by 50:1 during a batch. Heat transfer rates can vary in a continuous process, but rarely to this degree.

Edge-of-the-envelope issues. Measurement devices often exhibit unusual behavior at the lower and upper ends of the measurement range. Control valves behave strangely when operated at small openings (or "on their seat"). At large openings, oversized control valves cease to affect the controlled variable, or have such a small effect that control is not possible. Should limiting conditions within the process be encountered, the affected control loop ceases to be effective.

Control of Batch Processes, First Edition. Cecil L. Smith.
© 2014 John Wiley & Sons, Inc. Published 2014 by John Wiley & Sons, Inc.

This discussion assumes some familiarity with control concepts such as cascade, split range, overrides, and so on. A general treatment of these topics is available in an accompanying book [1].

This discussion is largely directed to the following two control issues that are especially important in many batch processes:

Accuracy of material charges. This is especially critical when two or more feeds must be in the stoichiometric ratios required by the reaction chemistry. Any error in the amount fed of any of these materials means that some unreacted material will remain, usually with negative consequences on the final product.

Temperature control. Reaction chemistry is strongly affected by temperature, leading to demanding requirements on controlling reactor temperatures. Issues arise for both control and measurement, but in addition, issues arise for the equipment being used to add or remove heat from the reactor. Multiple heating and cooling modes are very common in batch reactors. Transitions between modes occur during each batch, so smooth transitions are essential.

3.1. LOOPS THAT OPERATE INTERMITTENTLY

Except during start-up and shutdown, most loops in a continuous process operate 24/7. In a batch facility, loops such as those on the steam boiler in the utility area also operate 24/7. But where batch operations are conducted, most loops operate only during certain portions of the batch cycle. Such loops must start up smoothly, shut down smoothly, and not do strange things when they are not in use. This section will use a simple flow control loop as the basis for the discussion, but similar issues can arise for other loops.

Many flow loops in batch processes only operate for relatively short periods of time. A flow loop on a feed to the batch only operates when the feed is in progress. At other times, the flow is completely stopped, usually by a block valve. In essence, such flow loops operate intermittently. This is rarely the case in continuous processes.

The piping and instrumentation (P&I) diagram in Figure 3.1 illustrates a typical flow loop in a batch plant. To ensure that the flow is stopped, a common practice is to install both a control valve and a block valve. Figure 3.1 suggests

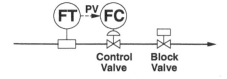

Figure 3.1. P&I diagram for flow control loop and block valve.

a separate block valve for each control valve, but this is not necessarily the case. For example, the various feeds to a vessel can enter through a header arrangement, with individual block valves for each feed and a common control valve. There are also "gimmicks" that permit a single physical valve to appear to the process controls as both a control valve and a block valve.

3.1.1. Zero Flow

With the block valve closed, the actual flow must be zero (assuming the valve does not leak). But will the measured value for the flow be zero? Flow transmitters require zero and span adjustments. Ideally, the zero should conform exactly to a flow of zero. But in practice, some error is always associated with the zero adjustment. Smart transmitters are far better than conventional transmitters in this regard, with many providing automatic zero adjustment. Consequently, the error is much smaller, but a zero flow is not assured to give a flow measurement of exactly zero.

Some transmitters provide a feature that forces the flow measurement to zero if the sensed value of the flow is less than some threshold (sometimes called a deadband). But as discussed shortly, adverse consequences arise if this feature is used inappropriately.

One of the arguments for using digital sensors is that zero is truly zero. If the flow is zero, the measured value will definitely be zero. But depending on the principle on which the measurement device is based, the measured value can be zero when fluid is flowing through the meter. The vortex shedding flow meter is a digital meter that counts vortices as they are formed and shed by the vortex shedder. As a valve closes and the flow decreases, the flow regime eventually transitions from turbulent to laminar. In the laminar flow regime, no vortices are formed, so the measured flow will be zero. That is, fluid flowing through the meter in laminar flow will result in a measured value of zero. Meter manufacturers state a minimum flow that can be sensed by the meter, and this must be taken seriously.

3.1.2. Stopping the Flow

With the flow controller in automatic, a value of zero for the flow set point should result in a fully closed control valve and a flow rate of zero. Sounds logical, and also sounds simple. However, this requires that the following two conditions be met:

1. The value for the controller output must be driven to its minimum possible value.
2. The minimum possible value for the controller output must cause the control valve to fully close.

Assuring these requires that issues be addressed for the flow transmitter and the final control element.

3.1.3. Final Control Element Issues

Most digital implementations of the PID block are configured so that the output is the percentage opening for the control valve. When configuring a PID block, the following are separately specified:

Output range (lower range value M_{LRV} and upper range value M_{URV}). For controllers whose output is a control valve opening, the output range is 0–100%. The control calculations are based on the output span $M_{URV} - M_{LRV}$. The units for the controller gain K_C are normally %/%, or more specifically, (% of output span)/(% of PV span).

Output limits (lower output limit M_{LOL} and upper output limit M_{UOL}). The value for the controller output is restricted by these values. Some, but not all, digital implementations permit the upper output limit to exceed the upper range value (provides an overrange) and the lower output limit to be less than the lower range value (provides an underrange). For example, the output limits might be specified as −5% to provide an underrange and 105% to provide an overrange. The default values are normally the maximum permitted underrange and overrange, with the user restricted to values for the output limits that are these or lesser values.

Why provide an underrange or overrange? Ideally, the control valve zero and span are perfectly adjusted so that 0% translates to a fully closed valve and 100% to a fully open valve. But the real world is not perfect, so one has to expect small errors in these adjustments. These are addressed as follows:

Underrange and overrange provided. An output value such as −5% should result in a fully closed control valve; an output value of 105% should result in a fully open valve. That is, the provided underrange and overrange exceed reasonable errors in the zero and span adjustments, thus assuring that the process controls can drive the control valve fully closed and fully open.

No underrange or overrange provided. To assure that an output of 0% will fully close the valve, the zero adjustment of the output range at the valve is set so that the valve is fully closed at an output value slightly above 0%. That is, instead of adjusting the control valve to operate between 0% and 100%, it is intentionally adjusted to operate between 2% and 100%. The controller is permitted to reduce the output value to 0%, which assures that the process controls can fully close the control valve. If desired, similar adjustments can be made to assure that the valve is fully open when the output is 100%.

In practice, these issues are generally most serious for valve fully closed. If a valve is not quite fully open, it is unlikely that anyone will notice. But a valve that is not quite fully closed is much like a leaking valve, and the consequences usually attract attention.

3.1.4. Flow Measurement Issues

What are the consequences of errors in the zero adjustment for the flow measurement? The two possibilities are as follows:

Measured value is zero when actual flow is slightly positive. Specifying zero for the flow set point would cause the flow controller to attempt to slightly open the control valve to attain the small flow necessary to give a measured value of zero for the flow. But if the control valve cannot be precisely positioned to such a small opening, the result is a cycle with the valve alternating between fully closed and its minimum opening.

Measured value is positive when actual flow is zero. This discussion assumes the integral mode is used in the flow controller, as is almost always the case. Consider the following conditions:

1. An actual flow of zero gives a measured flow that is slightly positive.
2. Zero is specified for the flow set point.

The flow controller will decrease its output to its minimum possible value, either 0% or something less depending on the underrange provided for control outputs. Assuming the zero adjustment on the control valve is properly set, this will fully close the control valve.

The latter is preferable to the former.

Prior to the advent of digital controls, a common practice in batch plants was to intentionally set the zero for flow measurements so that a zero flow gave a slightly positive measured value for the flow. One consequence is that a small positive flow would be indicated on the operator displays. This can be avoided by specifying a small deadband so that the flow is indicated as zero whenever the measured value for the flow is less than the deadband. But there is a precaution—the deadband must not be applied to the measured variable (PV) input to the flow controller. If the deadband is applied, the flow controller will not drive its output to its minimum possible value. A zero value for the set point and a zero value for the measured value translates to a control error of zero, so the integral mode ceases to change the controller output.

3.1.5. Discrete Logic

With conventional controls, practices such as intentionally biasing span adjustments were the simplest means to assure that a set point of zero resulted in a fully closed control valve. Although such practices have not entirely disappeared, digital controls provide preferable alternatives.

Most digital implementations of the PID control equation provide a feature known as output tracking (or something equivalent). Output tracking involves two additional inputs to the PID control block:

Value for output tracking (MNI). This is an "analog" value that is used for the output of the controller when output tracking is enabled. Herein, this input will be designated as **MNI**.

Enable output tracking (TRKMN). This is a discrete input that if true enables output tracking. Herein this input will be designated as **TRKMN**. When the **TRKMN** input is true, the controller does the following:

1. It forces the control output to the value of input MNI.
2. It suspends the PID control calculations, even if the loop is in the auto mode.
3. It executes the bumpless transfer calculations. This assures that the return to automatic control will be smooth or bumpless when **TRKMN** becomes false.

Most digital systems also permit the current value of the set point of a PID controller to be retrieved and used as an input to other functions. Herein this output will be designated as **SP**. The set point specified by the process operator is often called the local set point (**LSP**). In cascade configurations, the set point supplied by an external block is an input designated as the remote set point (**RSP**). The value of output **SP** is the current value of the set point, regardless of its source.

To force the control valve to close on a zero value of the flow set point, the inputs for output tracking must be configured as follows:

MNI. Input is the value for the output to the control valve when output tracking is enabled (**TRKMN** is true). The appropriate value is usually the lower output limit M_{LOL}.

TRKMN. Input is true when the set point is less than some minimum value, say F_{MIN}. An appropriate value for F_{MIN} is the minimum nonzero measurable flow.

The logic to force the control valve to close on a value of zero for the set point is not normally included in the P&I diagrams such as in Figure 3.1. Instead, the logic is included in more detailed drawings such as the control logic diagram illustrated in Figure 3.2. Alternatively, the logic can be expressed by simple relationships such as the following (**FCTAG** is the tag name of the flow controller):

$$\textbf{FCTAG.MNI} = \textbf{M}_{\textbf{LOL}}$$

$$\textbf{FCTAG.TRKMN} = (\textbf{FCTAG.SP} < \textbf{F}_{\textbf{MIN}})$$

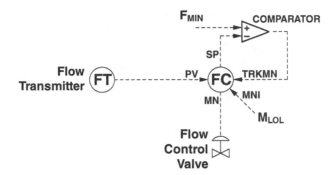

Figure 3.2. Control logic diagram for logic to close control valve on a set point of zero.

The comparison involves analog values, so whether the relational operator in the expression for **TRKMN** is "less than" (as in the aforementioned expression) or "less than or equal" is immaterial.

3.1.6. Windup in Flow Controller

Whenever the output of a PID controller has no effect on its measured variable, the integral mode within the PID controller will drive the controller output to either the maximum permitted value (the upper output limit) or the minimum permitted value (the lower output limit). Consider the following situation:

1. Block valve is closed.
2. Flow controller is in automatic with a nonzero set point.

With the block valve closed, opening the control valve has no effect on the flow. As the measured value for the flow is zero (or nearly zero), the controller drives the controller output to the upper output limit M_{UOL} (the control valve is fully open). The windup protection incorporated into the PID control equation is invoked when the controller output attains M_{UOL}, so the controller output does not increase beyond M_{UOL}.

Even though the control valve is fully open, no adverse consequences occur—the block valve is closed, and there is no flow through the valve. But consequences arise the instant the block valve is opened for the next transfer. As a result, the flow increases abruptly, which is undesirable and could have consequences. The flow controller quickly recovers, but the transition from no flow control to automatic flow control is certainly not smooth or bumpless.

A PID controller should not be performing control calculations whenever the conditions for windup are present. The following two approaches suspend the PID control calculations:

Switch the controller to manual. Also set the controller output to the minimum permitted value, specifically, the lower output limit M_{LOL}.

Enable output tracking. This suspends the PID calculations even if the controller is in auto. Furthermore, the bumpless transfer calculations are performed so that the control calculations can be resumed when output tracking is no longer imposed.

Usually the latter is the simplest to implement.

Let **RCVTAG** be the tag name of the block valve. Output tracking can be initiated on either of the following signals:

RCVTAG.ST. State of the block valve. Assume state 0 means valve is closed and state 1 means valve is open. Output tracking is to be enabled if the block valve is closed, as expressed by the following statement:

$$FCTAG.TRKMN = !RCVTAG.ST$$

RCVTAG.ZS0. Input from the limit switch that indicates that valve is fully closed. State 1 means that the valve is fully closed; state 0 means that the valve is not fully closed. Output tracking is to be enabled if the block valve is fully closed, as expressed by the following logic statement:

$$FCTAG.TRKMN = RCVTAG.ZS0$$

When output tracking is enabled, the output to the control valve must be the minimum possible value, as expressed by the following statement:

$$FCTAG.MNI = M_{LOL}$$

Using **RCVTAG.ST** as the basis for initiating tracking is usually preferable over using **RCVTAG.ZS0**. A subsequent chapter discusses the use of a discrete device driver for block valves equipped with one or more limit switches. To avoid disruptions in production operations resulting from a faulty limit switch, most discrete device drivers permit a limit switch to be "ignored," which means that the state of input **RCVTAG.ZS0** is to be "ignored" when the discrete device driver determines the state of **RCVTAG.ST**. Such an "ignore" is automatically incorporated into the tracking logic when based on **RCVTAG.ST**, but not when based on **RCVTAG.ZS0**.

3.2. EMPTYING A VESSEL

Figure 3.3 illustrates a reactor with three co-feeds. The product recipe stipulates the following:

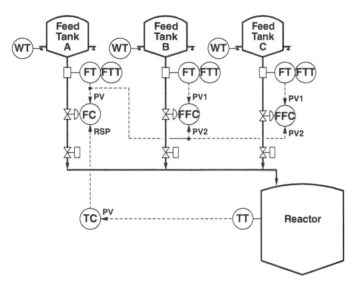

Figure 3.3. Reactor with three co-feeds.

1. The total amount of material A to be fed (feed A is the master flow)
2. The flow ratios for B-to-A and C-to-A.

From these, the total amounts of B and C can be computed.
    The following arrangements are encountered:

1. Add feed A at a constant rate. The feed rates for B and C are also constant. In manual operations, this approach is the norm. But as a process
   is automated, the incentive to shorten the feed time justifies operating
   at or near the highest permissible feed rate at all times.
2. Add feed A at a preprogrammed rate (e.g., change the flow rate of feed
   A based on time since start of the feed). Although the feed rates for B
   and C could also be preprogrammed, ratio control logic is usually
   preferred.
3. Adjust the flow of feed A to maintain the reaction temperature at the
   desired value. When heat transfer limitations impose a maximum on the
   feed rates, this usually gives the shortest feed time. Figure 3.3 illustrates
   the control logic for this approach. The reactor temperature controller
   adjusts the set point to the flow controller for feed A. The two FFC
   blocks are flow-to-flow controllers that maintain a specified ratio of the
   controlled flow PV1 (the flow through the control valve) to the wild flow
   PV2, which for both FFC blocks is the flow rate of A.

Figure 3.3 also provides flow totalizers (the FFT blocks) for each of the feed
flows.

When the reaction is stoichiometric, transferring the correct total amount for each feed is crucial. Small errors in the flow ratios normally have no perceptible effect on the final product. But if the total amount of any reactant is fed in excess, some unreacted material will remain in the final product.

The simplest approach is as follows:

1. Maintain the proper ratios of B-to-A and C-to-A.
2. Terminate all feeds when the flow total for feed A is attained.

However, this does not assure that the required amounts of B and C have been fed.

3.2.1. Feed Tank

Figure 3.3 provides a feed tank for each feed. Depending on the application, one or more of the feeds may be mixtures that must be prepared in advance. Two possibilities are the following:

1. Fill the feed tank with a sufficient quantity of the mixture so that several product batches can be manufactured. This approach is acceptable only if the mixture in the feed tank is stable, that is, does not degrade in some manner over time.
2. Fill the feed tank with the quantity of the mixture required for an individual product batch. Emulsions are usually prepared individually for each product batch.

The ensuing discussion applies to the latter case.

The following approach seems simple: start with an empty feed tank, fill with the required materials, transfer to the reactor, and make sure the tank is empty at the end. Just fill the tank with the appropriate materials and make sure that all is transferred to the reactor.

Sounds simple, but unfortunately a couple of issues arise:

1. How does one determine that the vessel is completely empty?
2. What issues arise with regard to the transfer piping?

3.2.2. Ascertaining That a Vessel Is Empty

Prior to automation, operators did this visually. For closed vessels, a nozzle can be outfitted with a sight glass and a tank light.

Conceptually the tank empty indication could be based on a continuous level or weight transmitter. However, both have issues with the zero. That is, zero does not exactly correspond to tank empty.

Weight transmitter. The weight of the empty vessel is its tare weight. Many weight transmitters are configured to indicate the weight of the contents only. This is done my subtracting the tare weight from the current weight of tank plus contents. Before starting to fill the vessel, the weight measurement can be "zeroed," which basically instructs the weight measurement to set the tare weight to the current weight, which should be that of the vessel only. But as noted in the previous chapter, installations of load cells on process vessels always involve compromises, the consequence being that the weight transmitter is unlikely to indicate zero when the vessel is emptied upon termination of the transfer.

Level transmitter. While the issues depend on the type of level transmitter, issues seem to arise for all types. When based on pressure measurements, the contents of the vessel at the lower range value of the pressure measurement depend on the location of the lower pressure tap. For radar instruments, the beam path determines the minimum level that can be sensed. However, the exposed bottom of a tank is unlikely to reflect the beam in the same manner as the liquid surface, which causes the measured value of the level to be invalid due to loss of the reflection.

Since the requirement is to detect vessel empty, an on–off indication is satisfactory. Level switches are based on a variety of technologies, the common ones being vibrating element, ultrasonic, and capacitance. Level switches are commonly installed in vessels to detect an abnormally low level. But as customarily installed, this does not correspond to vessel empty. One could propose to install the level switch at some location in the discharge piping. However, the piping in most batch facilities is rather small, which complicates this approach.

3.2.3. Driving Force for Fluid Flow

All of the following approaches are encountered:

1. Gravity flow
2. Vacuum in destination vessel
3. Pressurized source vessel
4. Transfer pump, either centrifugal or positive displacement.

For each of these, the coriolis meter is usually the preference. For gravity flow and also for vacuum in the destination vessel, the available pressure drop is limited, so the size of the vibrating tube must be larger than normally recommended, resulting in some sacrifice on the accuracy of the measured flow.

An advantage of the coriolis meter is its capability to sense the density of the fluid within the meter. This permits the transfer to be terminated on an empty coriolis meter. The piping arrangement must assure that the meter is empty of liquid at the end of the transfer. However, the arrangement must also assure that no air pockets can be present within the meter during the transfer.

3.2.4. Transfer Piping

With the objective of transferring the appropriate quantity of fluid, attention must be directed to the contents of the transfer piping, which usually means either of the following:

1. The transfer piping must be completely full before and after the transfer.
2. The transfer piping must be completely empty before and after the transfer.

When the transfer piping must be completely empty, the concept of "vessel empty" really means "vessel and transfer piping empty."

The best way to assure that the transfer piping is empty is to follow the transfer with a purge. Pressurized air (or suitable inert gas) can be used to "blow the line dry." This is a bane for turbine meter bearings—the rotational speed is high, and the bearings are not lubricated. Fortunately, such purges do not degrade other types of meters, including the coriolis meter.

Especially for pumped systems, purging with a liquid is common. However, issues such as the following arise:

1. Accurately metering the purge fluid is essential.
2. The transfer piping must be completely full of the purge fluid before and after the transfer.
3. The purge fluid enters the batch. Some enters at the start of the transfer, but some also enters at the end of the transfer.

Many batches begin with adding a sufficient quantity of some material to provide the "heel" so that the agitator can be started. One possibility is to use this material for the purge fluid. The heel can be shorted by the amount of fluid required for the purges (when the same material is also added later in the batch, one of the subsequent transfers can be shorted instead).

The total quantity of this material added to the batch is maintained constant, but the time that it is added changes slightly. The potential consequences include the following:

1. The heel will be smaller and possibly not sufficient to allow the agitator to be started.
2. The characteristics of the product could be affected.

For small flush volumes, neither is likely. But as noted in the introductory chapter, the effect of any change, no matter how small, on product characteristics must always be assessed.

3.3. TERMINATING A CO-FEED

The ensuing discussion applies to the case where the priority is to feed the specified total amount of each of the co-feeds. Options for terminating a co-feed include the following:

1. Terminate based on the flow totalizer (the FFT block in Figure 3.3) for that feed.
2. Terminate based on a load cell on the feed tank.
3. Assure that a feed tank is empty before and after the transfer.

Combinations are also possible—using the same approach on every co-feed is not necessary. However, each co-feed must be terminated such that the specified total is transferred to the reactor.

In a perfect world, each co-feed would terminate at exactly the same time. In practice, this does not occur. With the ratio configuration illustrated in Figure 3.3, the possibilities are as follows:

1. A co-feed stops before the master feed stops. This has no effect on the master feed or any other co-feed.
2. The master feed stops before all co-feeds have stopped. Assuming the flow meter for the master feed indicates a zero flow, this stops all co-feeds. Any that have not previously stopped will be stopped prematurely.

The latter observation applies to the ratio configuration where each co-feed is ratioed to the measured flow for the master feed, as is the case in Figure 3.3.

3.3.1. Ratio to Master Flow Set Point

The control configuration in Figure 3.4 is identical to that in Figure 3.3 but with one exception—each co-feed is ratioed to the master feed set point instead of the measured value of the master feed flow. The advantage of this configuration is that stopping the master feed does not stop the co-feeds, provided the original value of the set point for the master feed is retained when the master feed flow is stopped.

Implementing the configuration in Figure 3.4 is quite easy in digital controls. Most digital implementations use attributes to designate the possible values that can be retrieved from a PID control block. If **FCMASTER** is the tag name of the flow controller for the master feed, the following values can be retrieved:

FCMASTER.PV. Current measured value for the master feed flow. This should be the same as the output of the master feed flow transmitter. Either can be used to implement the configuration in Figure 3.3.

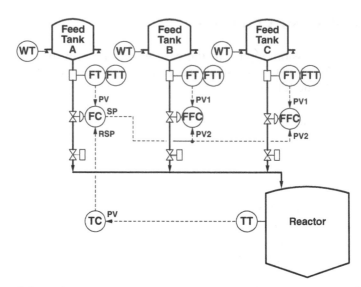

Figure 3.4. Ratio configuration based on flow set point instead of actual flow.

FCMASTER.SP. Set point for the master feed flow controller. If the mode
is automatic, this is the local set point. If the mode is remote or cascade,
this is the remote set point, which is provided by an input (often designated **RSP**) to the PID block.

Changing form the configuration in Figure 3.3 to the configuration in Figure
3.4 requires only one change to the ratio controllers: for the wild flow
input to each flow-to-flow controller, configure **FCMASTER.SP** instead of
FCMASTER.PV (or the output of the master feed flow transmitter).
   The configuration in Figure 3.4 is equivalent to the configuration in Figure
3.3 provided the master feed flow controller always maintains the master feed
flow close to its set point. For some applications, this is not assured, the usual
case being that the master feed flow controller fully opens its valve, but the
flow is still less than the set point. The consequences depend on the
configuration:

Ratio to measured value of master flow (Figure 3.3). The flow of the
   co-feeds will be reduced accordingly, maintaining the specified flow
   ratios.
Ratio to set point for master flow (Figure 3.4). The flow of the co-feeds is
   not reduced, which means the co-feeds are being added in excess.

The configuration in Figure 3.3 will reduce the co-feeds accordingly, but the
configuration in Figure 3.4 will not. Consequently, the actual ratio of each
co-feed flow to the master feed flow will be higher than desired.

3.3.2. Terminating Master Flow But Not Co-feed Flows

Maintaining the proper ratio of the co-feeds to the master feed is desirable throughout the batch. But in most cases, small errors usually have little effect on the batch. Near the end of the feeds, several possibilities exist:

1. Set the flow set points for the co-feed flows to fixed values.
2. Impose a minimum on each co-feed flow.
3. Set the flow on which the flow ratios are based to a fixed value.

In most cases, the difference in behavior is so small that it has no perceptible effect on the batch.

The required logic can be implemented in two ways:

1. In the sequence logic
2. In the regulatory control configuration.

Figure 3.5 illustrates an example of the latter approach. Two additional function blocks have been added:

Single-stage cutoff block (CUT). The purpose of this block is to provide a minimum value of the master feed flow near the end of the batch. In the configuration in Figure 3.5, the input to the cutoff block is the measured value of the weight (or level) of the feed tank for the master flow (an alternative approach is to base the logic on the value from the feed flow totalizer). The relationship for the single-stage cutoff block is as follows:

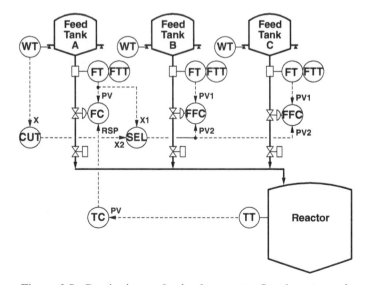

Figure 3.5. Continuing co-feeds after master flow has stopped.

$$F_{MIN} = \begin{cases} F_1 & \text{if } W \geq W_C \\ F_0 & \text{if } W < W_C, \end{cases}$$

where

W = Measured value of tank weight (or level)
W_C = Cutoff value for tank weight
F_0 = Value for cutoff block output if $W < W_C$
F_1 = Value for cutoff block output if $W \geq W_C$
F_{MIN} = Output of cutoff block

Coefficients W_C, F_0, and F_1 are configuration parameters for the cutoff block. For this application, F_1 must be zero.

Selector or auctioneer block (SEL). The output of this block is the larger of the following two values:

1. The measured value F of the master flow
2. The output F_{MIN} of the cutoff block.

If $W \geq W_C$, the output will be the measured value of the master flow. If $W < W_C$, the output will be the larger of F_0 and the measured value of the master flow.

The objective of the configuration in Figure 3.5 is to not stop a co-feed should the master flow stop before the co-feed stops.

Other configurations that accomplish this objective are certainly possible.

3.3.3. Cross-Limiters

The most common application of this technology is on combustion processes. The normal control logic should assure the proper air-to-fuel ratio. Deviations from the desired value lead to the following consequences:

Air flow is excessive for current fuel flow. The consequence is a loss of efficiency—the excess air flow increases the heat loss with the stack gasses.

Fuel flow is excessive for current air flow. Numerous consequences (such as a fire in the stack) are possible, most of which are bad.

Cross-limiters are installed primarily when a potential consequence is any type of hazardous situation. Cross-limiters are rarely justified on nonhazardous consequences (such as a loss in efficiency). These will be detected through other measurements (high stack oxygen levels), performance calculations, or other.

The objective of co-feeds is that materials are reacted in approximately the same ratio as they are fed. Should this not occur, possible consequences include the following:

1. Any material that is fed faster than it reacts builds up within the reactor.
2. Should a sufficient quantity of this material build up within the reactor, some reaction could begin to occur rapidly.
3. Especially for exothermic reactions, pressure increases are likely, possibly causing the reactor to be overpressurized.

When such hazards arise, cross-limiters could be considered. But in batch facilities, the following simpler approach is appealing:

1. Provide logic to detect when the excess of one of the feeds exceeds a specified value.
2. Stop the feeds and take any other appropriate actions (maybe impose maximum cooling).
3. Alert operations personnel.

The consequences of stopping a batch process are usually much less than stopping a continuous process.

3.4. ADJUSTING RATIO TARGETS

For the reactor with three co-feeds illustrated in Figure 3.3, the total amount of each co-feed that is fed to the reactor can be determined from two inputs:

Feed tank weight. Herein this is assumed to be a load cell. The reading from the weight transmitter at the start of the transfer is subtracted from the weight at the end of the transfer to give the amount transferred.

Flow measurement. A flow totalizer is provided for the flow of each co-feed. Prior to a transfer, the totalizer is reset to zero. After the transfer is completed, the value of the flow total is the amount transferred. Assume that the flow meter is a coriolis meter.

These will not agree perfectly, but they should be close.

Controls for the co-feeds in Figure 3.3 must address two issues:

Total amount. Being the most crucial issue, the total amount should be based on the most accurate measurement, either the vessel weight or the totalized flow.

Flow ratios. These can only the based on the flow measurements.

The accuracy of load cells versus coriolis meters was discussed in a previous chapter. The consequences of the conclusion as to which is most accurate will now be examined.

3.4.1. Interval for Taking Corrective Actions

In most co-feed applications, assessing the degree of agreement (or disagreement) after the co-feeds have been stopped is sufficient. But as tolerances become tighter, corrective actions are required while the co-feeds are in progress. Normally these are taken on certain intervals, two options being as follows:

Fixed time interval. The ratios are adjusted at times Δt_W, $2\Delta t_W$, $3\Delta t_W$, and so on, with time 0 being the start of the co-feeds.

Fixed quantity of feed A (the master flow). The ratios are adjusted upon feeding amounts ΔW_A, $2\Delta W_A$, $3\Delta W_A$, and so on.

The relationship between Δt_W and ΔW_A depends on the flow rate for feed A. Only for a constant feed rate are they algebraically related.

The interval must be sufficiently long that the noise in the weight measurements does not significantly affect the computations of the amount of each material that is transferred to the reactor over that interval.

The amount fed can be based either on the weight measurements or the flow totals. The following notation will be used for materials A and B:

$W_{WA,i}$	At the end of interval i, the total amount of feed A transferred to the reactor computed from the weight measurements.
$W_{WB,i}$	At the end of interval i, the total amount of feed B transferred to the reactor computed from the weight measurements.
$W_{FA,i}$	At the end of interval i, the total amount of feed A transferred to the reactor computed from the flow totals.
$W_{FB,i}$	At the end of interval i, the total amount of feed B transferred to the reactor computed from the flow totals.
$R_{WBA,i}$	Ratio of feed B to feed A for interval i, computed from weight measurements.
R_{WBASP}	Target for the ratio R_{WBA}. This value is obtained from the product recipe and is the same for every interval.
$R_{FBA,i}$	Ratio of feed B to feed A for interval i, computed from flow totals.
$R_{FBASP,i}$	Target for the ratio R_{FBA} for interval i.

As the notation and computations for material C are analogous to those for material B, only those for material B will be presented.

3.4.2. Flow Meter Deemed to Be Most Accurate

Terminating the feeds should be based on the flow totals. The flow ratio control is also based on the measured flows. Both are based on the most accurate measurement. The weight measurement is not used within the control logic.

Ideally, the feeds should terminate at exactly the same time. However, the ratio controls are not perfect, causing the feeds to terminate at slightly different times. In most applications, the discrepancies have no consequences. Often they occur during start-up of the co-feeds, resulting in an initial error that remains for the duration of the co-feed.

If necessary, logic can be implemented to compensate for such errors. For these computations, the total amount of each material to feed will be represented as follows:

$W_{FA,T}$ Total amount of A to feed. This value is usually explicitly specified in the product recipe.

$W_{FB,T}$ Total amount of B to feed. This value is the product of $W_{WA,T}$ and R_{WBASP}, as defined previously.

Both are based on flow totals.

As discussed earlier, adjustments will be made at the end of specified intervals, either based on Δt_W or ΔW_A. Except for the interval starting at time 0, the following computations will be performed:

1. Compute the remaining amounts to be fed for each material:

   $W_{FA,T} - W_{FA,i}$ Amount of A remaining to be fed.

   $W_{FB,T} - W_{FB,i}$ Amount of B remaining to be fed.

2. For the next interval, compute the set point for the flow-to-flow controller to feed the remaining material in this ratio:

$$R_{FBASP,i+1} = \frac{W_{FB,T} - W_{FB,i}}{W_{FA,T} - W_{FA,i}}$$

If the errors largely occur during start-up, the ratios computed by this logic will change on the first interval, but thereafter should approach the values for the ratios computed from the target feed amounts. If the ratios approach a different value, a source of error must originate from the performance of the ratio controls. As the same flow measurements are the basis for all of the control logic, the source of this error should be investigated.

A potential source of more significant errors is that the target flow for a co-feed cannot be attained. The controller fully opens the control valve, giving the maximum possible flow. The aforementioned logic will not compensate for this problem. For this co-feed, the logic will compute ever-increasing values for the ratio coefficient, which results in larger targets for the co-feed. If the present target cannot be attained, imposing a larger target is unproductive.

The only solution is to slow the feed rates for all feeds to the point that all can be maintained at their respective targets. This is the purpose of cross-limiters as routinely applied in combustion control applications. But as previously noted, cross-limiters are normally installed only when a hazardous situation could arise.

In batch applications, the most common approach is for operations personnel to resolve the issue. Usually the current batch is completed using a reduced target for the master flow. If the flow for the problem co-feed is being restricted in some manner (e.g., by a plugged filter), this problem must be corrected. Otherwise, the target for the master flow as specified in the master product recipe must be reduced.

3.4.3. Weight Measurement Deemed to Be Most Accurate

Terminating the feeds should be based on the weight measurements. The flow ratio control provided by the flow-to-flow controllers can only be based on the less accurate measured flows. The discrepancies between the values have the following consequences:

1. The feeds terminate at different times.
2. Materials are not fed in the exact ratio computed form the weight measurements.

In most cases, both differences are small and have little or no impact on the product batch.

Since the feeds are terminated based on the weight measurements, the flow totalizers illustrated in Figure 3.3 are not always necessary. But by retaining the flow totalizers, their values can be used as follows:

1. Compared with the values from the weight measurements, with operations personnel alerted on any discrepancy greater than a specified tolerance.
2. Used in computations whose objective is to feed materials in the proper ratio based on weight measurements instead of flow measurements.

Should logic with the objective of achieving ratios that are in close agreement with the weight measurements be deemed necessary, the following two approaches can be considered:

Adjust ratio targets. This is similar to the logic presented previously, but one aspect is different. The discrepancies between the weight measurements and the flow measurements usually require a flow ratio that is different from that computed from the target amounts.
Apply correction factors to the measured values for the flows. This has to be applied to all flows, including the master flow. Basically, the idea is to adjust the measured value of the flow to obtain a "believed" value of the flow that is consistent with the weight measurements.

The objective of terminating the feeds at the same time can also be incorporated into either.

3.4.4. Compensating Ratio Targets

The objective of the ensuing discussion is to adjust the set points for the flow-to-flow controllers such that the ratios computed from the weight measurements are the values specified by the product recipe. Only the computations for feed B will be formulated. However, the same is required for feed C.

The objective of the following formulation is to adjust the flow-to-flow controller set point R_{FBASP} so that the desired value for R_{WBA} is maintained throughout the feed. The value of R_{FBASP} will be adjusted on interval Δt_W or ΔW_A.

The general approach will be as follows:

1. For the first interval, specify the set point of the flow-to-flow controller to be the ratio of B to A from the product recipe.
2. At the end of the first interval, perform the following computations:
   (a) Based on the weight measurements, compute the amounts fed for each of the materials. From these, compute the actual ratio achieved over the time interval. This ratio is based on the weight measurements.
   (b) Based on the flow totals, compute the amounts fed for each of the materials. From these, compute the actual ratio achieved over the time interval. This ratio is based on the flow totals.
   (c) Compute a new value for the flow-to-flow controller set point so that for the next time interval, the ratio computed from the weight measurements is equal to the desired value.
3. Repeat step 2 at the end of each interval.

For material B, the two ratios are computed as follows:

$$R_{WBA,i} = \frac{W_{WB,i} - W_{WB,i-1}}{W_{WA,i} - W_{WA,i-1}} = \text{Actual ratio based on weight measurements.}$$

$$R_{FBA,i} = \frac{W_{FB,i} - W_{FB,i-1}}{W_{FA,i} - W_{FA,i-1}} = \text{Actual ratio based on flow totals.}$$

For flows, a common approach is to introduce a correction factor that will convert a flow measured on one basis to a flow measured on another basis. The following correction factor does the same for a flow ratio:

$f_{BA,i}$ Correction factor for converting the flow ratio from one based on weight measurements to one based on flow totals:

$$f_{BA,i} = \frac{R_{FBA,i}}{R_{WBA,i}}.$$

The desired ratio of B to A based on weight measurements is R_{WBASP}, which is a value from the product recipe and is the same for all intervals. Multiplying R_{WBASP} by the correction factor for interval i gives the target for the ratio based on the flow totals for the next interval, which is expressed by the following equation:

$$R_{FBASP,i+1} = f_{BA,i} \, R_{WBASP}.$$

This target is used as the set point for the flow-to-flow controller for feed B in Figure 3.3.

In practice, the adjustments should be relatively small, and should be limited to a narrow range around the desired ratio. Should larger adjustments be computed, the error in either a weight measurement or a flow measurement is excessive and needs the attention of operations and possibly maintenance personnel.

3.4.5. Flow Correction Factors

Sometimes called meter factors, the objective is to multiply each measured flow (including the master flow) by a factor to give a flow that is consistent with the weight measurements. In practice, only small corrections should be required, making coefficients significantly different from unity a cause for concern.

As discussed previously, the values for the correction factors are updated at specified intervals, the possibilities being a fixed time interval Δt_W or a fixed master feed quantity ΔW_A. Correction factors for feed flows are often referred to as "meter factors." For feeds A and B, the meter factors are computed as follows:

$$k_{A,i} = \frac{W_{WA,i} - W_{WA,i-1}}{W_{FA,i} - W_{FA,i-1}} = \text{Meter factor for feed A.}$$

$$k_{B,i} = \frac{W_{WB,i} - W_{WB,i-1}}{W_{FBA,i} - W_{FB,i-1}} = \text{Meter factor for feed B.}$$

The meter factor for feed C is computed in the same way. Multiplying the measured value of each flow by the respective meter factor gives a "believed" value for the flow that is consistent with the weight measurements.

As illustrated in Figure 3.6, a meter factor is applied to each flow. The flow inputs to the flow-to-flow (**FFC**) controllers are as follows:

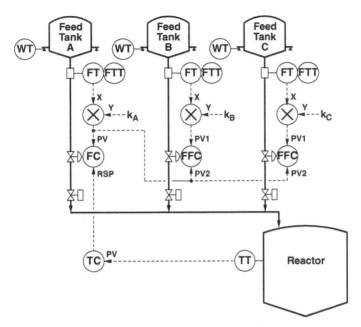

Figure 3.6. Correction factors for all flows.

Wild flow (PV1). The believed value for the master flow, which is the mea-
    sured value of the master flow multiplied by the correction factor for the
    master flow.

Controlled flow (PV2). The believed value for the co-feed flow, which is
    the measured value of the co-feed flow multiplied by the correction
    factor for that flow.

In Figure 3.6 the correction factor is also applied to the PV input to the master
flow controller. When the master flow set point is provided by a temperature
controller as in Figure 3.6, this is not necessary but does no harm. All meter
factors should be close to unity.

But if the desire is that the believed flow not be used in the temperature
control logic, the modified P&I diagram in Figure 3.7 can be used. The
meter factors are applied only to the flow inputs to the flow-to-flow
controller.

The correction factors must not be applied to the inputs to the flow totalizers. The flow totals used to compute the correction factors require totals for
the measured values of the flows.

If the value of a flow is used otherwise within the controls, a decision must
be made regarding using the measured value for the flow versus using the
believed value for the flow.

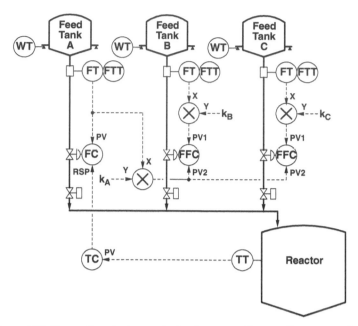

Figure 3.7. Correction factors only for inputs to flow-to-flow controllers.

3.4.6. Terminate All Feeds at Same Time

The objective of the logic discussed in the previous topic was for the ratio of feed B to feed A be as close as possible to R_{WBASP} on every interval. No attempt is made to compensate for errors in the ratio on previous intervals, so the feeds will not terminate at the same time.

The total amount of each material to feed will be represented as follows:

$W_{WA,T}$ Total amount of A to feed. This value is usually explicitly specified in the product recipe.

$W_{WB,T}$ Total amount of B to feed. This value is the product of $W_{WA,T}$ and R_{WBASP} as defined previously.

Both are based on weight measurements, not on the flow totals.

At the end of interval i, the amounts remaining to be fed are computed as follows:

$W_{WA,T} - W_{WA,i}$ Amount of A remaining to be fed.

$W_{WB,T} - W_{WB,i}$ Amount of B remaining to be fed.

To terminate at the same time, the remaining materials must be fed in this ratio, which is based on the weight measurements.

This ratio can be converted to a ratio based on flow totals by either of the following approaches:

1. Compute the correction factor $f_{BA,i}$ for the flow ratio as previously discussed. Then compute the target for the flow ratio of B to A as follows:

$$R_{FBASP,i+1} = f_{BA,i} \frac{W_{WB,T} - W_{WB,i}}{W_{WA,T} - W_{WA,i}}$$

2. Compute the meter factors for each flow as previously discussed. Then compute the target for the flow ratio of B to A as follows:

$$R_{FBASP,i+1} = \frac{(W_{WB,T} - W_{WB,i})}{(W_{WA,T} - W_{WA,i})}.$$

Since the meter factors are applied to the inputs to the flow-to-flow controller, the ratio based on weight can be used as the flow-to-flow controller set point.

A variation of this approach is for the objective to be to feed the appropriate amounts over the coming interval so that the total amounts fed at the end of that interval will be in the ratio specified in the product recipe. That is, the desire is to adjust the feeds for the coming interval so as to compensate for all previous deviations from the ratio specified in the product recipe. The formulation is much easier if the interval is based on a specified quantity of A, specifically, ΔW_A.

3.5. ATTAINING TEMPERATURE TARGET FOR THE HEEL

Many batch reactions begin with adding sufficient nonreacting materials to the vessel to provide a heel, which permits the agitator to be started. With no agitation, the following statements apply:

The heat transfer rate will be very small. Consequently, the heel must be nonreacting.

The vessel temperature measurement is meaningless. Even if the sensor is submerged, stratification is very likely. Once sufficient material is added to provide the heel, the agitator must be run for some period of time before the measurement of the vessel temperature attains a stable value.

The specifications for the heel always state the amount of each material to be added. Occasionally, the specifications include the heel temperature. Specifically, the stated value is the desired temperature of the heel within the vessel,

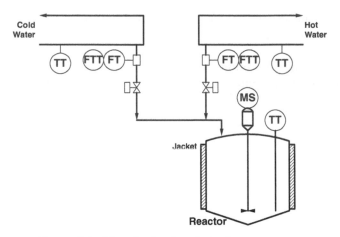

Figure 3.8. Charging heel to desired temperature.

which could be different from the temperature of the combined material fed to the vessel to provide the heel.

3.5.1. Mixing Hot and Cold Fluids

Figure 3.8 illustrates a vessel for which the heel consists entirely of water. Separate feeds are provided for hot water and cold water. Flow measurements and temperature measurements are provided for each. For the temperature measurement to be meaningful, the fluid must be flowing. In Figure 3.8, the temperature measurements are installed in the recirculation loops for the hot and cold fluids.

If the resulting heel temperature is not within the acceptable range, the temperature can be adjusted by heating or cooling via the jacket. However, this increases the time to produce the batch, effectively lowering plant productivity. This provides an incentive to get the right temperature when adding the materials for the heel.

The amounts of hot and cold water to mix to obtain a temperature equal to the heel temperature are computed from the following mass and energy balances:

$$M = M_{\mathrm{H}} + M_{\mathrm{C}}$$

$$M\,T = M_{\mathrm{H}}\,T_{\mathrm{H}} + M_{\mathrm{C}}\,T_{\mathrm{C}},$$

where

M = Mass for heel, kg
T = Temperature for heel, °C
M_{H} = Mass of hot fluid to be added, kg

T_H = Temperature of the hot fluid, °C
M_C = Mass of cold fluid to be added, kg
T_C = Temperature of the cold fluid, °C.

The specifications provide values for M and T. Values for T_H and T_C are available from the process measurements. The aforementioned two equations can be solved for M_H and M_C. Each flow is terminated when its flow totalizer attains the respective target.

This assures that the combined water flows for the heel will have the desired temperature. But once in the vessel, thermal equilibrium is established between the heel, the vessel, and whatever is in the jacket. In effect, a term for heat addition or loss should be included in the above energy balance. If this term is significant, the final heel temperature will be different from the temperature used to compute values for M_H and M_C from the aforementioned equations.

3.5.2. Contribution of Vessel and Jacket

The contribution of the vessel and jacket to the heel temperature depends on several factors:

1. The larger the heel, the less the vessel and jacket affect the heel temperature.
2. The vessel temperature depends on the conditions at the end of the previous batch. For vessels in which the same batch is run repeatedly, the effect on heel temperature will be very similar for each batch. That is, it will consistently raise or lower the heel temperature by some amount. If so, this can be incorporated into the computations of M_C and M_H. But suppose the process is operated 24 hours a day but only 5 days a week. The first batch Monday morning probably behaves differently from subsequent batches.
3. The contents of the jacket could depend on the previous batch. This depends on the nature of the heating media. If cooling was required, the jacket is filled with water, glycol, or something similar. If heating was required and heating is provided by condensing steam, the jacket may be empty.

3.5.3. Two-Stage Addition of the Heel

The ability to more consistently attain the desired heel temperature can be enhanced by the following approach:

1. Add sufficient water (mixed to the desired heel temperature) so that the agitator can be started.
2. Let the heel temperature attain equilibrium.

3. Compute the temperature at which the remaining water must be added to attain the desired heel temperature.

Unfortunately, heels are often sized to be the minimum amount required to start the agitator, making the previously mentioned approach unusable. The possibility of increasing the size of the heel depends on the nature of the batch.

Conceptually, this could be extended to a three-stage addition of the heel. This requires an even larger heel. Since thermal equilibrium must be established after each addition to obtain an accurate measurement of heel temperature, more time is also required. This reduces the advantages over using a smaller heel and adjusting the heel temperature using the jacket.

3.6. CHARACTERIZATION FUNCTIONS IN BATCH APPLICATIONS

Most continuous processes operate in a very narrow region, so the variation of the process characteristics is sufficiently small that the effect on the performance of the process controls is nominal. Being subjected to large changes in throughput, utility processes are an exception. For a steam boiler, the turndown ratio (maximum steam rate to minimum steam rate) is usually in the range of 4:1, and rarely exceeds 10:1. However, treatment facilities for wastewater do encounter turndown ratios exceeding 10:1.

To maintain the required performance over the range of steam flows specified by the turndown ratio for the boiler, logic beyond simple feedback control must be incorporated into the controls. During commissioning, tests are performed to demonstrate that the boiler will operate over the required range of steam flows. These and possibly other tests provide the data for constructing characterization functions that relate other flows (or possibly valve openings) to the steam flow or steam demand. The resulting characterization functions are incorporated into the controls to provide feedforward control action.

Without the characterization functions, the feedback controllers provide the primary response to both set point changes and disturbances, making the performance of these controllers crucial to control system performance. But by incorporating the characterization functions into the controls for a steam boiler, the role of the feedback controllers is to provide the feedback trims that compensate for the inevitable inaccuracies in the characterization functions. This makes the performance of the feedback controllers less important.

3.6.1. Throughput in a Batch Process

The throughput in many reaction processes is the heat transfer rate between reactor contents and jacket. For the moment, assume constant conditions for the heating media (e.g., constant steam supply pressure or hot oil supply temperature) or cooling media (e.g., constant cooling water supply temperature).

At a point in time, the conditions in the reacting media determine the heat transfer rate. But to achieve maximum production rates (shortest batch times), the process controls must adjust conditions in the reacting media so that the heat transfer rate is the maximum possible.

Developing characterization functions and controlling in a manner similar to a steam boiler could be considered, but there is a complication. Unlike steam flows, heat transfer rates cannot be directly sensed. Potentially, the heat transfer rate can be computed from measurements on the heating or cooling media. For cooling water or hot oil, the heat transfer rate is given by the sensible heat equation:

$$Q = F\ c_P\ (T_{OUT} - T_{IN}),$$

where

Q = Heat transfer rate from vessel to jacket
F = Flow of heating or cooling media
c_P = Heat capacity of heating or cooling media fluid
T_{OUT} = Exit temperature of heating or cooling media
T_{IN} = Inlet temperature of heating or cooling media.

The 50:1 turndown ratio presents problems at the extremes (the "edge of the envelope"):

Low end (small flow; high temperature rise). To measure the flow at the low end, a flow meter with a high turndown ratio is required. Even if the low flow can be measured, the accuracy of most flow meters is expressed as a percentage of the upper range value. On the low end, accuracy as a percentage of reading suffers.

High end (high flow; small temperature rise). Computing the temperature rise by subtracting two large numbers (the temperatures) to obtain a small one (the temperature rise) amplifies errors in the measured values of the temperatures. Using a differential temperature transmitter is recommended, but even so, any error in the temperature difference is amplified when multiplied by the large flow rate.

Such issues greatly complicate measuring the throughput, which in turn impedes the use of characterization functions in batch applications.

3.7. SCHEDULED TUNING IN BATCH APPLICATIONS

A batch process experiences changes in the process characteristics throughout the batch. If not addressed, these degrade the performance of the controls, and product quality suffers. Scheduled tuning is one possibility for addressing this issue.

3.7.1. Re-Tuning Controllers

In process applications, feedback control is synonymous with PID control. A critical component of commissioning any PID controller is tuning, which entails getting the characteristics of the controller "in tune with" the characteristics of the process. The PID controller is a linear controller and performs best in applications with reasonably constant process characteristics. In batch processes, this proves to be challenging—the characteristics of the process change throughout the batch.

Even in continuous processes, PID controllers must occasionally be re-tuned. The interval between re-tuning is measured in months or years, making manual procedures quite acceptable. But batch applications present the following obstacles:

1. Re-tuning would be required during each batch, and possibly at several times within the batch. Doing this manually is unrealistic.
2. Automatic tuners that require any form of a process test are unacceptable. The test(s) would have to be performed during each batch.
3. The automatic tuners that do not employ tests respond too slowly to track the changes in process characteristics during the batch.

3.7.2. Components of Scheduled Tuning

In most batch applications, the need to change controller tuning parameters as each batch proceeds is generally recognized. The customary approach is a technique known as scheduled tuning, the components of which are the following:

1. The envelope of process operations is divided into regions. More on this shortly.
2. The controller is tuned within each region and the results saved in a tuning table. A row is provided for each region; a column is provided for each tuning coefficient. From a process control systems perspective, the number of regions could be very large. However, the need to tune the controller within each region usually limits the size of the tuning table. Rarely are more than five regions encountered.
3. The scheduled tuning logic involves determining the current region of operation and copying the appropriate tuning parameters from the tuning table to the controller.

The possibilities for defining the regions include the following:

Time in batch. Usually measured as time since some significant event (such as inoculating a fermentation). Process conditions change as the batch proceeds. Instead of attempting to quantify the changes in some manner, the time in batch is used to define the regions for the tuning table.

Events within the batch. Some reactions consist of an initiation phase (the reaction is initiated at a low feed rate), the continuous feed phase (the feed rates are maintained at much higher rates until all is added), and a completion phase (the temperature is increased so as to drive the reactions to completion). The tuning table can contain tuning coefficients for each phase of the reaction.

Throughput. Instead of computing the heat transfer rate, the regions for the tuning table are defined based on the heating or cooling media valve opening or, if available, the media flow rate.

The challenge is to properly define the regions. If the process is deemed to be in a certain region, the process characteristics must be approximately the same so that the tuning coefficients from the tuning table give good loop performance. Otherwise, the controller must be conservatively tuned to the most demanding process characteristics encountered within that region. This degrades the performance of the controller and lessens the benefits of scheduled tuning.

3.7.3. Limits of Scheduled Tuning

As the batch progresses, some process characteristic can change to the extent that the PID controller is not capable of delivering the required control performance. Although most likely to arise as the process operating conditions approach the edge of the envelope, this can potentially occur under other conditions. If the controller cannot be successfully tuned within a region, scheduled tuning can only take actions such as switching the loop to manual.

Process sensitivity generally decreases as throughput increases. A turndown ratio of 50:1 on heat transfer means a low sensitivity at high heat transfer rates and a high sensitivity at low heat transfer rates. Changes in process sensitivity primarily affect the tuning parameter for the proportional mode. The controller gain or sensitivity is inversely proportional to the process gain or sensitivity. A large value for the controller gain is required at high heat transfer rates and a low value of the controller gain is required at low heat transfer rates.

For digital process controls, the limits on the value of the controller gain are often dictated by factors such as numerical formats. Imposing a format of XX.XX on the controller gain permits the controller gain to be as high as 100%/% (technically 99.99%/%) and as low as 0.01%/%. With the equation as usually written, specifying 0.00%/% for the controller gain makes no sense (the controller gain multiplies all terms in the control equation). If allowed, a controller gain of 0.00%/% imposes a slightly different control equation that provides integral-only control. This is not appropriate for temperature control.

In practice, such extreme values can rarely be used, the results being as follows:

Controller gain of 100%/%. Any noise or other imperfection in the measured value of the controlled variable is greatly amplified in the output of the controller.

Controller gain of 0.01%. The controller takes extremely small control actions, the result being equivalent to the loop being on manual. Control valves do not respond to changes below some threshold, so small changes have to accumulate before the valve moves. Variable speed drives respond to much smaller changes than control valves, but not to changes generated by a controller gain of 0.01%/%.

Such situations must be recognized so that repeated attempts at tuning are not undertaken. The alternatives are as follows:

Disable the controls. That is, switch the control loop to manual. But if the controlled variable is temperature, this will not be perceived as a "solution."

Change the control approach. For reactors with a continuous feed phase, consider applying maximum heating or cooling and control reactor temperature by varying the feed rate. Before making such a change, get the blessings of the product chemists. But if they object, make sure the plant manager knows that such an approach gives the maximum possible reaction rates, which shorten the time for making a batch. Unless the chemists' objections are convincing, plant managers have ways to get everybody on board, especially when there are orders than cannot be filled.

3.8. EDGE OF THE ENVELOPE

All processes are designed to operate over a range of conditions, which can be thought of as the operating envelope. The "edge of the envelope" consists of the boundaries imposed on process operations, the most common factors being the following:

Equipment limitations. The operating pressure must not exceed the pressure rating of vessels.

Product limitations. The operating temperature may not be below the freezing point of the process fluid, above the bubble point of the process fluid, above the temperature at which the product degrades rapidly, and so on.

Control limitations. Conditions that cause some component of the process controls to cease to function must be avoided.

The first two are generally well understood and can occur in both continuous and batch processes. The latter often results from some subtle characteristic of the process, and seems far more common in batch processes than continuous ones. This will be the focus of the following discussion.

3.8.1. Behavior at the Edge of the Envelope

Should the process controls attempt to operate the process outside the operating region, the result is that one or more loops do not deliver acceptable performance. The behavior is usually one of the following:

1. A cycle appears in both the controller output and the variable being controlled. Sometimes this is mistaken for a tuning problem, leading to repeated attempts to "tune out the cycle." Adjusting the tuning parameters affects the nature of the cycle, but can never eliminate it. Applying automatic tuners and/or tuning techniques also prove unsatisfactory — the cause of the cycle is not a problem with the tuning. However, if the objective is merely to keep the loop on automatic, reducing the controller gain eventually increases the period of the cycle to beyond the time to manufacture the batch. However, the controller responds so slowly that it is ineffective.

2. The controller drives its output to either the upper output limit or the lower output limit. With no remaining control action, the controlled variable either drops below or rises above its set point. Sometimes, this occurs well before the final control element attains its limiting value, giving the impression that a problem exists within the final control element. However, the culprit is a limiting condition being imposed within the process, but possibly made worse by two factors:

   (a) Oversizing of one or more components within the loop. Often this is the control valve, but could be some component within the process.

   (b) Windup resulting from the reset or integral action within the controller. A simple example pertaining to a flow loop was presented previously, but often the cause leading to windup is more subtle and needs further explanation.

3.8.2. Windup

Control loops rely on some relationship through which the changes in the controller output affect the process variable. When this cause-and-effect relationship breaks down, the following statement is valid:

Changes in the controller output have no effect on the variable being controlled.

In practice, "no effect" often actually means "insufficient for effective control." In most cases, the breakdown occurs gradually, but it can occur abruptly. In the earlier example of the flow controller, windup immediately commences on closing the block valve with the flow controller on automatic and not tracking.

The previously mentioned statement is the condition that permits windup to occur. The mechanism is as follows:

1. The process variable is above or below its set point, resulting in a control error.
2. The presence of this error causes the reset mode to increase or decrease the controller output. The direction depends on the sign of the control error and the controller action.
3. Changes in the controller output have no effect on the process variable, so the control error persists.
4. With a persistent control error, the reset mode continues to increase or decrease the controller output.
5. This continues until the controller output attains either the upper output limit or the lower output limit. The windup protection incorporated in the controller is invoked at either the upper output limit or the lower output limit.

Windup is associated with the reset mode and is sometimes referred to as "reset windup." A controller with no reset action will not go into windup.

Windup commences when changes in controller output cease to affect the measured variable. Windup protection needs to be invoked the instant that this occurs (or shortly thereafter). In an ideal world, the windup protection should be invoked at the limits of the output range (normally 0% and 100% for control valves). But due to errors in ranging the final control element, the output limits (such as -2% to 102%) provide a small overrange to assure that the controller can fully close and fully open the valve. The windup protection provided within the controller is invoked at the output limits, so a small degree of windup occurs. This is justified to assure that the control valve can be fully opened or fully closed.

3.8.3. Invoking Windup Protection Other Than at the Output Limits

Although the mechanisms vary from one supplier to another, all digital controls provide mechanisms to properly invoke windup protection. Doing this is subject to the following requirements:

1. Why the controller output ceases to affect the process variable must be understood. The loop functions properly under most conditions, so why it encounters problems under certain conditions can be subtle. Often the suspected culprit is either inappropriate tuning or a problem within the final control element. However, no progress will be made until the focus

is on the process—there is some aspect that causes the problem to arise under certain conditions. As experienced process control professionals like to tell new hires, "You have to understand the process." Good advice, but philosophical. Rarely is it followed up by explaining exactly how one goes about doing this.

2. The controls must be able to determine when such conditions arise. This may require process measurements that are not routinely installed, such as detecting when a steam trap is blocking the steam flow.

3. Windup prevention mechanisms (external reset, integral tracking, or inhibit increase/inhibit decrease) must be properly configured. Controllers automatically invoke windup protection at the output limits. To invoke windup protection earlier, the relevant features must be properly configured. The configuration issues are explained in an accompanying book [1], so they will not be explained here.

3.8.4. Recognizing Why Loops Cease to Function

When asked how to do it, the usual response is to give examples (or "war stories," if you prefer). The challenge is to translate these into a set of general principles that can be used as guides to the process analysis. Herein examples will be given for the following:

Problems resulting from how the final control element is used. The final control element is working properly, but it is being expected to do something that it cannot do. Such problems arise due to the following:

1. Flow through the control valve is zero.
2. Pressure drop across the control valve is zero. An oversized valve can contribute to this behavior.

Limits imposed within the process. Some characteristic of the process imposes a limit on process operations, which in turn causes the control valve to cease to affect the measured variable. Examples will be presented for the following:

1. Attempting to operate above a maximum imposed within the process
2. Attempting to operate below a minimum imposed within the process.

Unfortunately, this is not necessarily a complete list of situations that can cause changes in the controller output to cease to affect the measured variable.

3.9. NO FLOW THROUGH CONTROL VALVE

Figure 3.9 illustrates an example where a control valve is manipulated to control the upstream pressure P_1. The same configuration could be applied

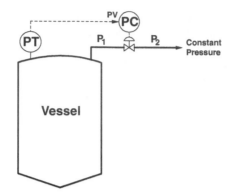

Figure 3.9. Pressure control.

to control the downstream pressure P_2. The action of the controller must be changed, but otherwise, the issues are the same.

3.9.1. Control Objective

If $P_{1,SP}$ is the target or set point for the upstream pressure P_1, the controller must adjust the valve opening until the following is achieved:

$$\Delta P_V = P_{1,SP} - P_2.$$

The pressure drop ΔP_V across the control valve is related to the flow F through the valve and the valve opening M by the following generic relationship:

$$\Delta P_V = f(F, M).$$

The relationship depends on the inherent valve characteristics and the nature of the flow system, specifically, the pressure drop across the valve relative to the pressure drop in the remainder of the flow system.

The nature of this relationship could lead to tuning issues. But even with the best tuning, problems always arise at the boundary condition or "edge of the envelope." Regardless of control valve and flow system characteristics, the following statement is true:

$$\text{If } F = 0, \Delta P_V = 0 \text{ for all values of } M.$$

Changing the valve opening M has no effect on the flow F through the control valve, and consequently no effect on the upstream pressure P_1.

In practice, this statement is never exactly true. Instead, as the flow through the valve decreases, the controller responds by decreasing the valve opening. This continues until the control valve is operating with a very small opening or "on its seat." All actuators have a limit to their ability to position the valve, resulting in a minimum valve opening that the actuator can sustain. Below this, the valve alternates between fully shut and some small opening, resulting in a cycle.

3.9.2. Vacuum Control

In batch facilities, such an issue can arise in controlling the pressure in vessels under vacuum. If the vessel in Figure 3.9 operates under vacuum, the pressure is being controlled via a valve in the piping to the vacuum equipment. As long as there is adequate flow through the control valve, this configuration works quite well. During the main portion of the batch, there are three sources of gases:

1. Gas dissolved in the feeds will be released.
2. Many complex reactions release small amounts of gas through side reactions.
3. Leaks are always present, despite efforts to prevent them.

At the end of the reaction phase of a batch, two sources disappear:

1. Feeds are stopped. No dissolved gas is released.
2. As reactants are depleted (sometimes to trace levels), all reaction rates drop to nearly zero. No gas is produced by side reactions.

The flow through the control valve is down to the leaks, resulting in a very small valve opening. The controller opens the valve slightly, causing the pressure to drop below the target. Once the controller closes the valve, the pressure slowly increases, eventually causing the controller to open the valve slightly. Once started, the resulting cycle repeats indefinitely.

3.9.3. Inert Gas Bleed

The configuration in Figure 3.10 assures that there is always some flow of gas through the control valve. A small flow of an acceptable inert gas (nitrogen, methane, carbon dioxide, or anything that does not react with materials within the process) is bled into the vessel. The consumption of inert gas is small, but assures that the flow through the control valve is sufficient for the pressure controller in Figure 3.10 to function properly.

The pressure controller will function properly provided the inert gas is admitted at any location upstream of the control valve. But in processes where oxygen is a bad actor, arguments can be made for admitting the inert gas into the vessel. As the inert gas flows through the vessel and into the vacuum system, some of the oxygen is swept along. In batch stills operating under vacuum, the inert gas can be admitted into the pot, causing it to flow up the column and into the vacuum system.

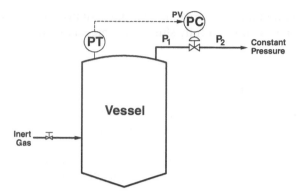

Figure 3.10. Providing an inert gas bleed.

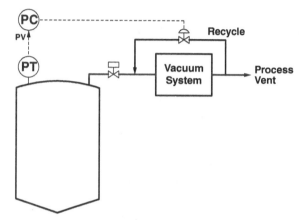

Figure 3.11. Pressure control using recycle around vacuum equipment.

A possible variation of the control configuration is the following:

1. Instead of installing the control valve in the vacuum line, install it in the line that admits the inert gas into the vessel.
2. Replace the control valve in the vacuum line with a block valve.

3.9.4. Recycle

For vacuum pump or ejector systems, the inert gas can be obtained by recycling gas from the first stage discharge. The control configuration in Figure 3.11 is the result of the following modifications:

1. The control valve in the line to the vacuum system is replaced by a block valve.

2. The recycled gas is admitted upstream of the block valve. But when recycled gas is used in lieu of a true inert gas, no advantages accrue from admitting the gas into the vessel itself (as in Figure 3.10).

3. The control valve is installed in the recycle line.

3.10. NO PRESSURE DROP ACROSS CONTROL VALVE

The effectiveness of any control valve depends on the pressure drop across the control valve relative to the pressure drop in the remaining part of the piping. Oversizing control valves is so common within the industry that one gets the impression that it has become a tradition. The valve sizing equations are well known, but the sizing calculations require values for available pressure drop, required maximum flow, and so on. Using conservative values leads to oversized valves. In flexible batch facilities, this is a major problem. The valve has to be sized for the most demanding product in the range of products currently being manufactured. The temptation to throw in a little extra for the product that might go into production next year is irresistible.

3.10.1. Flow through Control Valve

In some applications, the maximum flow through the control valve is determined entirely by process issues and is in no way dependent on valve sizing. A good example is the stream heating jacket in Figure 3.12. The maximum steam flow rate is determined by the maximum possible rate of heat transfer to the process, the equation being the following:

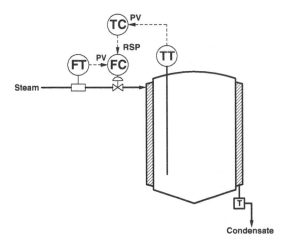

Figure 3.12. Vessel with steam-heated jacket.

$$Q_{MAX} = UA\,(T_S - T)$$

$$F_{MAX} = Q_{MAX}/\lambda,$$

where

Q_{MAX} = Maximum possible heat transfer rate from jacket to vessel
F_{MAX} = Maximum possible steam flow rate
U = Heat transfer coefficient
A = Heat transfer area
T_S = Saturation temperature of steam at steam supply pressure
T = Temperature of vessel contents
λ = Latent heat of vaporization for the steam.

Especially when the valve is oversized (as it typically is), this limit is approached. For large valve openings, the pressure of the condensing steam is very close to the steam supply pressure, which gives a very small pressure drop across the control valve. In this region, changing the valve opening has little effect on the condensing steam pressure, which has little effect on the condensing steam temperature and the heat transfer rate.

3.10.2. Windup Issues

Figure 3.12 illustrates a temperature-to-flow cascade for controlling the vessel temperature. Consider operating with the steam flow controller on local automatic, which means that the flow controller is using the local set point entered by the process operator instead of the remote set point supplied by the temperature controller. What happens when value for the local set point exceeds F_{MAX}? The flow controller fully opens the control valve on the steam, and the measured value of the steam flow approaches F_{MAX}.

If the temperature controller is provided a set point higher than can be obtained, a potential consequence is windup. The vessel temperature will be below its set point for a lengthy period of time. If the controller contains reset action (as most do), the temperature controller will eventually specify a steam flow set point that exceeds F_{MAX}. Whenever the temperature is below its set point, the reset action continues to increase the temperature controller output (steam flow set point) until the upper output limit is attained. At this point, further increases in the controller output are not permitted, and windup protection is invoked within the controller.

But for the control configuration in Figure 3.12, windup has already occurred. Once the remote set point (the output of the temperature controller) exceeds F_{MAX}, further increasing the remote set point has no effect on the steam flow, the heat transfer rate, and the vessel temperature. This is a condition for windup—changes in the output of the temperature controller (the remote set point for the flow controller) have no effect on its measured variable (the vessel temperature).

As normally configured, the upper range value for the output of the temperature controller in Figure 3.12 corresponds to the upper range value for the steam flow measurement. So that all possible steam flows can be measured, the upper range value for the steam flow measurement exceeds the value of F_{MAX}, at least in most cases. This permits the temperature controller to specify a steam flow set point above what can be obtained, which creates the potential for windup.

3.10.3. Windup Protection

The appropriate value for the upper output limit for the temperature controller is F_{MAX}, not the upper range value for the flow measurement as customarily configured. But doing so raises two issues:

1. The value of F_{MAX} is not a constant, but depends on conditions within the process.
2. A value cannot be computed for F_{MAX} because accurate values for coefficients such as U are not available.

In such situations, other approaches must be used to prevent windup. All digital control systems provide at least one of the following:

External reset. This applies to the reset feedback form of the PID, and was provided by conventional pneumatic and electronic controls. The current value of the steam flow is the external reset input to the temperature controller, as illustrated by the control logic diagram in Figure 3.13.

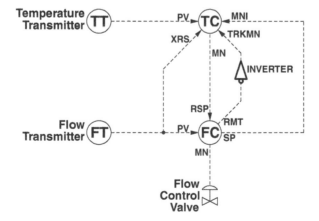

Figure 3.13. Control logic diagram for temperature-to-flow cascade with external reset.

Integral tracking. When the steam control valve is fully open, the current value of the steam flow is substituted for the output of the reset mode.

Inhibit increase/inhibit decrease. When the steam control valve is fully open, the temperature controller is not allowed to further increase its output.

Although not exactly the same, any one will prevent the windup. The specifics of each are presented in an accompanying book [1], so no further discussion will be presented here.

3.10.4. Advantage of the Cascade Configuration

Perhaps windup does not technically occur in the flow loop in the cascade configuration in Figure 3.12, but the behavior is similar to what one observes with windup. For a given batch, the heat transfer equations determine the maximum possible steam flow F_{MAX}. Assume that the control valve is an equal-percentage valve and, on the basis of this steam flow, is oversized by a factor of 4. Below a valve opening of 50%, changes in the control valve opening affect the steam flow and control is possible. But above 50%, the flow is "maxed out" at F_{MAX} or a value close to F_{MAX}. The effect of changes in the flow controller output has a small effect on the steam flow, but too small for flow control to be effective.

Once the temperature controller specifies a set point above F_{MAX}, the flow controller drives its output to the upper output limit. Flow dynamics are fast, so it will do this rather rapidly. What if the temperature controller then reduces its output to a value below F_{MAX}? The flow controller begins to reduce its output, but effective flow control does not commence until the controller output drops to 50% or less. Increasing the output of the flow controller from 50% to 100% is effectively windup. Before effective flow control is established, the flow controller has to reduce its output from 100% to 50%. In essence, the controller is unwinding. Flow loops are fast and will unwind relatively quickly, so the impact on the performance of the temperature controller is minimal.

What if the flow loop is removed? The temperature controller output is now setting the opening of the steam valve. The valve behavior is the same as before. Below an opening of 50%, the temperature can be controlled by manipulating the opening of the steam valve. But above an opening of 50%, changing the valve opening has insufficient effect on the steam flow and hence the temperature for the temperature controller to be effective.

If the temperature controller is given a set point that requires a steam flow greater than F_{MAX} to achieve, the temperature controller will drive its output to its upper output limit. Basically, the valve opening is increased from 50% to 100%, but far more slowly due to the slow dynamics of the temperature process. Controllers that wind up slowly also unwind slowly. If the temperature set point is subsequently lowered to a value that can be achieved with a steam flow less than F_{MAX}, the temperature controller will first slowly decrease its output from 100% to 50%. Only then will effective control of temperature resume.

In this regard, the temperature-to-flow cascade performs much better than simple temperature control. Due to the fast dynamics of the flow loop, the oversized valve has little impact on temperature control in the temperature-to-flow cascade. But in the simple feedback configuration, the slow temperature dynamics significantly delay the time required for effective temperature control to resume. This advantage of cascade is rarely included in a list of advantages for cascade, but it can be a significant advantage in certain applications in batch facilities.

3.11. ATTEMPTING TO OPERATE ABOVE A PROCESS-IMPOSED MAXIMUM

This example may appear similar to the previous example of the steam-heated jacket, but there is a difference. For the steam-heated jacket, attempting to operate above the maximum possible steam flow resulted in a very small pressure drop across the valve, causing the effect of changes in the valve opening on steam flow to be very small.

In the following example, the process imposes a maximum that is unrelated to the control valve. Changes in the control valve opening continue to affect the flow through the valve, but the nature of the process is such that changes in the flow have no effect on the variable being controlled.

The water-cooled vessel in Figure 3.14 will be used as an example where that can occur. The vessel in Figure 3.14 is similar to the one in Figure 3.12 except that the vessel is cooled with cooling water instead of heated with steam.

3.11.1. Maximum Cooling Rate

For the vessel in Figure 3.14, the maximum heat transfer rate is given by the following expression:

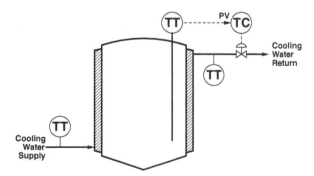

Figure 3.14. Vessel with water-cooled jacket.

$$Q_{MAX} = UA\,(T - T_{CWS}),$$

where

  Q_{MAX} = Maximum possible heat transfer rate from vessel to jacket
  U = Heat transfer coefficient
  A = Heat transfer area
  T_{CWS} = Cooling water supply temperature
  T = Temperature of vessel contents.

At very high cooling water flows, the temperature rise from cooling water supply to cooling water return is very small, meaning that the cooling water return temperature is essentially the same as the cooling water supply temperature.

Under these conditions, the ΔT for heat transfer is the vessel temperature T less the cooling water supply temperature T_{CWS}. At high cooling water flow rates, changes in the cooling water flow have no effect on the ΔT for heat transfer, and consequently no effect on the heat transfer rate or the vessel temperature. In this context, "no" actually means "negligible"—the effect is not exactly zero, but is far too small to be effective for control.

Oversizing is not limited to control valves. The ability to flow water through jackets such as in Figure 3.14 is often excessive, especially in batch facilities. Compared with other components of the manufacturing costs, cooling water is inexpensive and pumping costs are trivial. The temptation to oversize is irresistible, even by factors of 10 or more.

In this case, the problem is not with the control valve. Changing the valve opening affects the flow of cooling water through the control valve. However, changes in the cooling water flow have very little effect on the heat transfer rate.

3.11.2. "Edge of the Envelope" for Control with Cooling Water Flow

The situation described earlier only occurs at very high cooling water flow rates. At lower cooling water flow rates, changes in the cooling water flow affect the heat transfer rate, making it possible to control vessel temperature by varying the control valve opening and hence the cooling water flow.

The vessel temperature can be effectively controlled by manipulating the cooling water flow provided the heat transfer rate is less than about 90% of Q_{MAX}. The heat transfer rate can potentially be computed using the sensible heat equation for the cooling water, but a couple of issues arise:

1. A measurement is required for the cooling water flow.
2. Especially at high cooling water flows, the temperature rise from jacket inlet to outlet is small, necessitating a differential temperature measurement.

Fortunately, the limit can be expressed in terms of temperatures.

Removing heat from the vessel in Figure 3.14 involves two mechanisms:

Heat transfer. Heat must be transferred from the contents of the vessel to the jacket. At high cooling water flows, the following statements apply:
1. The heat removal rate is limited by the heat transfer rate.
2. The maximum ΔT for heat transfer is $T - T_{CWS}$.
3. Vessel temperature cannot be controlled by manipulating cooling water flow (or valve opening).

Sensible heat. Heat must be removed from the jacket via the sensible heat of the cooling water. At low cooling water flows, the following statements apply:
1. The heat removal rate is limited by the sensible heat relationships.
2. The maximum temperature rise for the cooling water is $T - T_{CWS}$.
3. Vessel temperature can be controlled by manipulating cooling water flow (or valve opening).

The transition from sensible heat limited to heat transfer limited can be characterized by the following parameter:

$$\frac{\text{Temperature rise of cooling media}}{\text{Maximum temperature difference for heat transfer}} = \frac{T_{CWR} - T_{CWS}}{T - T_{CWS}}.$$

A heat transfer rate of 90% of Q_{MAX} translates approximately to the following condition:

$$\frac{T_{CWR} - T_{CWS}}{T - T_{CWS}} \cong 0.2.$$

If the value of the ratio is less than 0.2, changing the cooling water flow has little effect on the heat transfer rate, so the vessel temperature cannot be effectively controlled by manipulating the cooling water flow.

The transition from sensible heat limited to heat transfer limited is gradual, and certainly does not occur abruptly at a value of 0.2 for the aforementioned ratio. In practice, slightly larger values are sometimes recommended for the transition.

3.11.3. Override Control

When the ability to flow cooling water through the jacket is grossly oversized (such as by a factor of 10), fully opening the control valve allows so much water to flow through the jacket that other cooling water users are "starved." One approach for avoiding this is to impose a minimum value on the

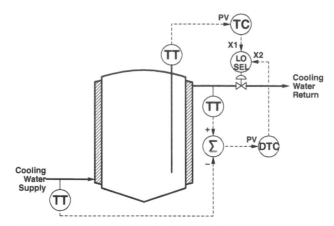

Figure 3.15. Vessel temperature control with override on cooling water temperature rise.

temperature rise of the cooling water, that is, $T_{CWR} - T_{CWS}$. The appropriate value can be computed from the ratio presented earlier:

$$[T_{CWR} - T_{CWS}]_{MIN} = 0.2\,(T - T_{CWS}).$$

In a sense, the value of $[T_{CWR} - T_{CWS}]_{MIN}$ can be viewed as an override on the reactor temperature control. The override configuration in Figure 3.15 is based on the following approach:

Primary objective. Adjust the cooling water valve opening (or cooling water flow) so as to maintain the reactor temperature at its set point. The reactor temperature controller in Figure 3.15 provides this function. The set point is the desired value for the reactor temperature.

Override. If the temperature rise is equal to or less than $[T_{CWR} - T_{CWS}]_{MIN}$, then adjust the cooling water valve opening (or cooling water flow) such that the cooling water temperature rise equals $[T_{CWR} - T_{CWS}]_{MIN}$. The cooling water rise temperature controller in Figure 3.15 provides this function. The set point is $[T_{CWR} - T_{CWS}]_{MIN}$.

The selector in Figure 3.15 switches between these two by selecting the larger of the two controller outputs. Such selectors must be configured properly to avoid windup. This and other aspects of override control are presented in an accompanying book [1].

3.11.4. Vessel Temperature to Cooling Water Temperature Rise Cascade

For most applications of this type, the cascade configuration in Figure 3.16 is much simpler. The two loops are configured as follows:

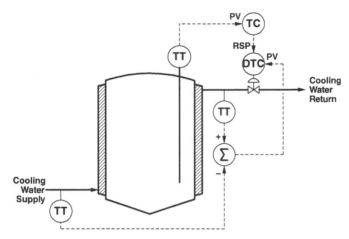

Figure 3.16. Vessel temperature to cooling water temperature rise cascade.

Outer loop. Vessel temperature controller (**TC**).

Inner loop. Cooling water temperature rise controller (**DTC**).

The output of the vessel temperature controller is the set point for the cooling water temperature rise controller.

To avoid excessively large cooling water flows, a minimum must be imposed on the set point for the cooling water temperature rise controller. This limit must be imposed in a manner such that windup protection is invoked in the vessel temperature controller should the controller output attain the minimum for the cooling water rise. There are two options for imposing the limit:

Lower output limit for vessel temperature controller. Windup protection is initiated in the vessel temperature controller when the lower output limit is attained.

Set point for temperature rise controller. Most implementations provide a lower set point limit as part of the configuration, but if necessary, a limiter block can be inserted into the cascade configuration. However, additional configuration may be required to initiate windup protection in the vessel temperature controller when such limits are being imposed.

3.11.5. Vessel Temperature to Cooling Water Return Temperature Cascade

The control configuration in Figure 3.16 responds very effectively to disturbances in the cooling water supply temperature. But in most applications, this is not necessary. Seasonal variations occur in the cooling water supply temperature, but during a batch, the cooling water supply temperature usually varies little, if any. In such cases, the cascade configuration in Figure 3.17 is equivalent to the cascade configuration in Figure 3.16.

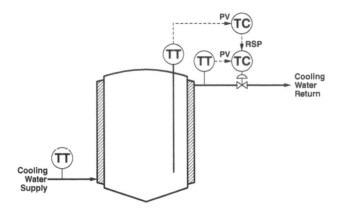

Figure 3.17. Vessel temperature to cooling water return temperature cascade.

A minimum must be imposed on the set point for the cooling water return temperature controller, again in such a manner that windup protection is invoked in the vessel temperature controller should its output attain the minimum for the cooling water return temperature set point.

3.11.6. Consequences for Vessel Temperature

The configurations in Figure 3.15, Figure 3.16, and Figure 3.17 have one aspect in common. Once the cooling water temperature rise attains its minimum value, the vessel temperature is no longer being controlled.

Although blame is often directed at the process controls, the origin of the problem is on the process side. Attempts are being made to operate the process in such a manner that the required heat transfer rate is above the maximum possible heat transfer rate.

Objections are often raised to imposing limits on the cooling water flow. If the ability to flow water through the jacket is significantly oversized, the minimum of the cooling water temperature rise will occur at cooling water valve openings of 50% and even less. Those without a good process background have the impression that significant cooling capacity remains, which is unfortunately not the case. The process is being operated at about 90% of the maximum possible heat transfer rate.

If large cooling water flows are not starving other vessels for cooling water, limiting the cooling water flow is a difficult argument to make. Relative to the margins for most products of batch facilities, the cost of cooling water is trivial. One possibility is to fully open the cooling water valve and let the vessel temperature "float," that is, allow the vessel temperature to seek an equilibrium. However, people such as product chemists will not accept this for a temperature loop.

For heat transfer rates above about 90% of maximum, the sensitivity of vessel temperature to changes in cooling water valve opening (or flow) is too small for effective control. To operate at full cooling, the following approach is required:

1. Fully open the cooling water valve.
2. Find another way to control vessel temperature.

For some reacting systems two or more feeds are added continuously over some portion of the batch. Usually, the heat transfer rate is proportional to the feed rates, which permits the vessel temperature to be controlled by manipulating the feed rates. Unfortunately, such options are not always available.

3.12. ATTEMPTING TO OPERATE BELOW A PROCESS-IMPOSED MINIMUM

In these cases, the minimum is entirely unrelated to the control valve. Changes in the control valve opening affect the flow through the valve. But because of process considerations, flows below a certain rate cannot be sustained. Attempts by the controller to attain lower flow rates result in cycling.

The vessel with the steam-heated jacket in Figure 3.12 will exhibit this type of behavior. The contributing factors are as follows:

1. The minimum pressure of the condensing steam in the jacket is atmospheric pressure. A pressure above atmospheric is required for the condensate to flow out of the jacket.
2. At atmospheric pressure, the saturation temperature of steam is 100°C, which is the minimum possible condensing steam temperature.
3. If the vessel temperature T is less than 100°C, the minimum driving force for heat transfer is 100°C − T.
4. The minimum heat transfer rate Q_{MIN} is given by the following expression:

$$Q_{MIN} = UA\,(100°C - T),$$

   where U is the heat transfer coefficient, and A is the heat transfer area.
5. The minimum steam flow FMIN that can be sustained is Q_{MIN}/λ, where λ is the latent heat of vaporization of steam at atmospheric pressure.

3.12.1. Cycling

If the set point for the steam flow controller in Figure 3.12 is less than F_{MIN}, cycling will commence, the components being as follows:

1. A steam flow less than F_{MIN} causes the pressure of the condensing steam to drop, eventually attaining atmospheric.

2. With the condensing steam pressure at atmospheric, there is no driving force for the condensate to flow out of the jacket. Hence, condensate accumulates within the jacket.

3. The effective heat transfer area is the surface area exposed to condensing steam. The submerged surface area cools the condensate, with little contribution to heat transfer to the vessel. Consequently, the effective heat transfer area decreases as the jacket fills with condensate.

4. The reduced heat transfer rate causes the vessel temperature to decrease. In response, the vessel temperature controller increases the steam flow set point, eventually attaining a value that exceeds F_{MIN}.

5. The higher steam flow increases the pressure of the condensing steam, eventually attaining pressures above atmospheric pressure.

6. A condensing steam pressure above atmospheric blows all of the condensate out of the jacket, exposing all available heat transfer surface to condensing steam.

7. The increased heat transfer rate causes the vessel temperature to increase. In response, the vessel temperature controller decreases the steam flow set point, eventually attaining a value that is less than F_{MIN}.

8. The cycle repeats from step 1.

For the temperature-to-flow cascade in Figure 3.12, the cycle has a short period that reflects the dynamics of the flow loop, which are relatively fast. If the flow loop is removed, a cycle still occurs, but with a much longer period that reflects the dynamics of the temperature loop.

For the jacket arrangement in Figure 3.12, varying the steam flow affects the condensing steam pressure, and consequently, the condensing steam temperature and the heat transfer rate. But if the vessel temperature is less than 100°C, this cause-and-effect relationship only works for condensing steam pressures above atmospheric. To attain heat transfer rates less than the rate corresponding to a condensing steam temperature of 100°C, a fundamentally different approach is required.

3.12.2. Alternate Configuration

The jacket arrangement in Figure 3.18 differs only in that the control valve is on the condensate instead of the steam supply. The temperature driving force is always the saturation temperature at the steam supply pressure less the vessel temperature. The heat transfer rate is affected through changes in the effective heat transfer area. The jacket is partially filled with condensate at all times. Varying the condensate flow affects the quantity of condensate retained within the jacket, which affects the heat transfer area exposed to condensing steam and consequently the heat transfer rate.

At least theoretically, the jacket could be completely filled with condensate, giving no heat transfer area and no heat transfer. Consequently, the minimum heat transfer rate that can be sustained is essentially zero.

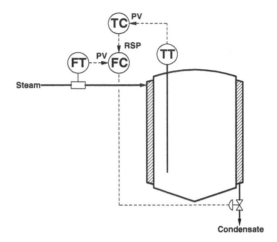

Figure 3.18. Vessel with steam-heated jacket, control valve on condensate.

With the valve on the condensate, the process responds more slowly to a change in the control valve opening. However, production vessels usually provide the desired 5:1 separation of dynamics between the inner loop (the flow loop) and outer loop (the temperature loop) of the cascade in Figure 3.18.

The slower dynamics do impact the tuning of the flow controller. Although often not clearly stated, the customary tuning coefficients for a flow controller apply when the measured flow is the flow through the control valve. In Figure 3.18, the measured flow is the steam flow; the flow through the control valve is the condensate flow. Consequently, the customary tuning coefficients (such as a controller gain of 0.2%/% and a reset time of 3 seconds) are unlikely to be satisfactory for the flow controller.

3.12.3. Blowing Steam

A separate issue is that the control valve in Figure 3.18 can "blow steam" at high heat transfer rates. As the condensate valve opens, less condensate is retained in the jacket. In most cases, the jacket will completely empty of condensate before the control valve is fully open. If so, steam flows through the control valve, which is not acceptable.

A simple solution is to install a steam trap upstream of the control valve. When the jacket contains condensate, it flows freely through the trap, with the condensate flow rate determined by the control valve opening. When the jacket is fully drained of condensate, the trap allows the condensate to pass but blocks the steam.

Although installing the trap is a simple solution, the potential for windup arises. When the jacket is drained of condensate and the trap is blocking the steam, the control valve opening has no effect on the steam flow. This is the condition for windup. Windup protection should be invoked the instant the trap begins to block the steam flow. Digital process controls provide a variety of features pertaining to windup protection, but no signal is available to the process controls indicating that the steam trap is blocking steam, and consequently, windup protection needs to be invoked.

In production facilities, condensate must usually be returned to the boiler house and not dumped to a drain. One advantage of the configuration in Figure 3.18 is that the full steam supply pressure is available for condensate return. Designs involving condensate pots will prevent the jacket from "blowing steam," but these will not be presented here.

3.13. JACKET SWITCHING

Jacketed vessels in batch facilities are often capable of providing either heating or cooling. Some provide multiple modes of cooling, giving options such as the following:

1. Heating with steam
2. Cooling with tower water ($\sim 20°C$)
3. Cooling with chilled water ($\sim 5°C$)
4. Cooling with refrigerated glycol ($< 0°C$).

This raises issues pertaining to executing the transitions between the various heating/cooling modes, which herein will be referred to as "jacket switching."

When choosing the heat transfer equipment associated with the jacket and the control logic required for executing transitions, a major consideration pertains to the requirements for the process. These requirements generally fall into the following two categories:

The required heating/cooling mode is known for each step or phase of the batch processing. The jacket can be switched to the required heating/cooling mode at the start of each phase. Making such transitions is the subject of this section.

The required heating/cooling mode must be determined from conditions within the process. Logic within the process controls must execute the switch at whatever time it is required. Furthermore, the transition must be smooth and executed expeditiously, especially when the switch occurs with reacting materials in the vessel. Making these transitions is the subject of the sections that follow.

3.13.1. Jacket with Four Heating/Cooling Modes

Figure 3.19 illustrates a jacket capable of being switched between the four heating/cooling modes described previously. Within a company, the equipment

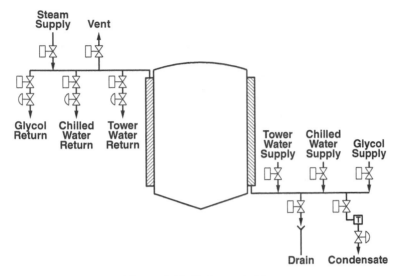

Figure 3.19. Jacket switching.

configuration for such jackets is usually very similar from one installation to the next. This is not the case between companies, the reasons including the following:

1. Requirements of the products being manufactured
2. Characteristics of the heat transfer equipment installed in the facility
3. Cost
4. Individual preferences, some dating from years past.

The configuration in Figure 3.19 reflects the following considerations:

1. The tower water, chilled water, and glycol enter the jacket at the bottom, flow upward through the jacket, and return from the top of the jacket. For each service, block valves are installed on both the supply and the return. From a control perspective, the control valve can be located on either the supply or the return. Figure 3.19 locates the control valves on the returns, which gives the highest pressure within the jacket.
2. Steam is supplied to the top of the jacket; the condensate is removed at the bottom of the jacket. Block valves are located on both supply and return. For small vessels, the steam condensate sometimes flows to a drain, but for production scale vessels, the condensate is usually returned to the boiler house, as illustrated in Figure 3.19. The control valve for the steam can be either on the supply or on the condensate. The issues were briefly discussed in the previous section of this chapter, and are the same as for the steam heater presented in an accompanying book [2]. For batch

applications, the ability to sustain small heat transfer rates is desirable. To do so usually requires that the steam control valve be located on the condensate return as in Figure 3.19. To assure that the jacket cannot "blow steam" into the condensate return, a steam trap is located upstream of the condensate control valve.

3. At times the jacket will be pressurized with steam. Before switching to a cooling mode, the jacket must be depressurized. The vent valve in Figure 3.19 is installed for this purpose.

4. Before switching from any cooling mode to steam heating, the jacket must be emptied of its current contents. For the configuration in Figure 3.19, the contents could be forced into the appropriate cooling media supply by pressurizing the jacket with either compressed air or possibly steam. The drain included in Figure 3.19 can be used to empty the jacket of tower or chilled water, but doing so for glycol is preferably avoided. One possibility is to dump tower and chilled water to the drain, but force the glycol into the glycol supply (or provide additional valve and piping so that the glycol can be forced into the glycol return).

5. The jacket configuration in Figure 3.19 is a once-through jacket. Once-through jackets present a nonuniform ΔT between the jacket and the vessel contents, which has a negative impact on the manufacture of some products. The uniform ΔT provided by a recirculating jacket is preferable. The equipment configuration is different, but the issues associated with jacket switching are largely the same.

The various alternatives associated with jacket configurations have advantages and disadvantages, but most considerations ultimately come down to money. That such considerations lead to different jacket configurations should not be surprising. Additional piping, valves, instrumentation, and so on, could be installed to mitigate most issues, but these cost money to install and maintain.

3.13.2. Transitions

For the jacket in Figure 3.19, there are 12 possible transitions. To switch to a given service (such as tower water), three transitions are possible:

1. Steam to tower water
2. Chilled water to tower water
3. Glycol to tower water.

For each of the four possible services, there are three possible transitions, giving a total of 12.

The easiest transitions to execute are between cooling with tower water and cooling with chilled water. For both modes, the jacket is filled with water. Only the following steps are required:

1. Close the supply and return block valves for the cooling media currently in service.
2. Open the supply and return block valves for the cooling media to be used in the future.

But when either steam or glycol is involved, the contents of the jacket must change. For the jacket in Figure 3.19, this is the case in 10 of the 12 possible transitions.

For all transitions that require a change in the jacket contents, consider structuring the logic using the following transitions:

1. From the current heating/cooling media to the "Jacket Empty" state. There are four such transitions, but the transitions for tower water and chilled water are the same.
2. From the "Jacket Empty" state to the desired heating/cooling media. There are four such transitions, but the transitions for tower water, chilled water, and glycol are the same except for the block valves on the media supply and return.

The "Jacket Empty" state may not be exactly what the name implies. For a jacket initially filled with water, the possibilities for emptying the jacket include the following:

Drain the jacket by gravity. An atmospheric vent valve is required. The plant vent system is at a slight negative pressure and would admit contaminants into the heating and cooling media.

Use compressed air to force the water from the jacket. Figure 3.19 does not provide such a capability, but it could be installed. This would shorten the time required to drain the jacket, but would leave the jacket pressurized with air. Once fully drained, releasing to a vent leaves a jacked filled with air at or near atmospheric pressure. This raises the following issues:

1. When filling with a liquid, a "bleeder valve" must be installed to release the air but not the liquid. But if installed, the "bleeder valve must be blocked off during steam heating.
2. When filling with steam, the jacket must be purged with steam to reduce the noncondensable components to an acceptable level.

Use steam to force the water from the jacket. This definitely makes sense when switching to steam heating—the need to purge the jacket is eliminated. When switching to cooling, the steam in the jacket would be initially vented, but the remaining steam is either released through the bleeder valve or condenses into the cooling media.

When the jacket must be drained and refilled, the time to switch the heating/cooling mode is lost production time. Obviously, the time must not be excessive. But in addition, the jacket is not available for control purposes. To avoid excessive temperature excursions, jacket switching can only be undertaken when no significant reactions are occurring within the vessel.

3.13.3. Instrumentation Considerations

Implementing the logic for the transitions can be approached from two extremes:

1. Use timing for each action required to make the transition. For example, to empty the jacket of water, open the steam supply valve, open the drain valve, wait for a specified length of time, and assume that the jacket is empty. The consequences of errors in the time are as follows:

   Time is on the short side. Some liquid remains in the jacket, but the consequences of a small amount may be acceptable.

   Time is on the long side. The jacket will "blow steam" into the drain. If this undesirable, a steam trap can be installed upstream of the drain valve.

2. Terminate each action based on a continuous or discrete measurement. Terminating emptying the jacket of water requires that a detector of some type be installed to indicate that the jacket contains no liquid. The issues are similar to those discussed in a previous chapter for emptying a feed tank.

The first approach is more common, mainly due to the "keep it simple stupid" philosophy coupled with the costs of installing and maintaining the additional measurements required for the second approach.

Controller tuning means adjusting the tuning parameters for proportional, integral, and derivative such that the controller is "in tune" with the characteristics of the process. The logic for the transitions is tuned in a similar fashion. The various times are adjusted to match the time required by the process to complete the various actions. The adjustments are not perfect, and will need revisiting from time to time. But provided the consequences of small errors in a time are tolerable, basing the logic on timing is usually the preferred approach. When implementing the logic, provisions must be made for adjusting the various times used by the logic.

3.13.4. Implementing the Logic for Jacket Switching

As will be examined in detail in a later chapter, the logic for automating batch processes can be divided into two categories:

1. Product-specific, equipment independent
2. Equipment-specific, product independent.

The logic for jacket switching falls into the latter category. For manufacturing hexamethylchickenwire, switching from steam heating to tower water cooling is executed in exactly the same manner as for other products.

The logic within the product recipe initiates switching from one heating/cooling mode to another. In most cases, the product recipe logic cannot proceed until the jacket has been switched, so it must be somehow informed that the jacket has been switched to the appropriate mode. Basically, the product recipe logic instructs the jacket switching logic to "do this, tell me when it is completed, and don't bother me with the details."

Maintaining the jacket switching logic separate from the product recipe logic has another advantage. From time to time, modifications are made in the heat transfer equipment, in the measurements installed on the jacket, and so on. Usually, the objective is to provide a faster switch, to achieve a smoother switch, to mitigate consequences associated with the switch, and so on. Such modifications will require changes to the jacket switching logic, but usually not to the product recipes. Only when a new feature is added (such as another heating/cooling mode) is the interface between the product recipe logic affected by the modifications.

In most applications, jacket switching logic is one component of the jacket control logic. A common requirement is "ramp and soak." Integrating such capabilities into the jacket control logic permits the product recipe logic to direct the jacket control logic to heat the vessel from the current temperature to a target temperature at a specified rate (or in a specified time), and then maintain the temperature at the target for a specified duration of time (the "soak time"). The temperature target, the ramp rate (or ramp time), and the soak time vary from one product to another. But the detailed actions required to achieve the objectives rarely depend in any way on the product being manufactured.

3.14. SMOOTH TRANSITIONS BETWEEN HEATING AND ONE COOLING MODE

The following exothermic reaction is an example of an application that requires smooth transitions between the available heating/cooling modes:

1. At the start, cooling is required to remove the heat released by the reaction.
2. After one or more reactants have been largely consumed, the rate of the reaction decreases to the point that heating is required to maintain the desired reactor temperature (heat from the reactions is less than heat loss to ambient).

The requirements include the following:

1. The transition from cooling to heating occurs during the reaction phase, but not at a time that can be precisely predicted in advance.

2. The transition must be initiated based on process conditions. When cooling is no longer required to maintain the reactor temperature, the transition to heating must ensue.
3. As a reaction is occurring, the transition must be executed quickly. Not enough time is available for emptying and filling the jacket.
4. Temperature excursions, if any, must be small. Product chemists are always concerned about the impact of temperature excursions on reacting systems.

Such transitions are commonly initiated by a control logic known as split range, which is described in an accompanying book [1]. However, to achieve a smooth transition, an appropriate equipment configuration for the jacket is also required.

3.14.1. Type of Jacket

The heat transfer rate in some batch processes changes dramatically from one part of the batch to another. Low heat transfer rates normally result in low flow rates for the heating and/or cooling media. The implications are as follows:

Once-through jacket. Low cooling/heating media flow rates translate to low flow rates through the jacket, which can lead to poor distribution, rapid changes in the temperatures within the jacket, a nonuniform ΔT between jacket and vessel contents, and so on.

Recirculating jacket. The flow within the recirculation loop is high at all times. The temperature rise from jacket inlet to jacket outlet should be small (5°C or less). The ΔT between the jacket and the vessel contents is essentially uniform, even when the heating/cooling media flows are low.

For applications with widely varying heat transfer requirements, the recirculating jacket is preferred. But it is not always installed.

3.14.2. Steam Heating; One Cooling Mode

The jacket configuration in Figure 3.20 provides only two modes:

1. Heating with steam
2. Cooling with tower water.

A smooth transition from cooling to heating (and also from heating to cooling) is possible with this jacket configuration. The jacket is filled with water at all times and is equipped with a recirculating pump. The behavior is as follows:

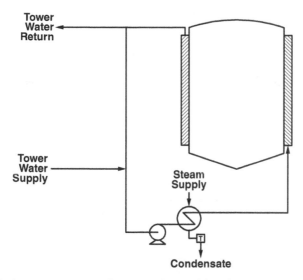

Figure 3.20. Jacket arrangement for smooth transition between steam heating and tower water cooling.

Cooling mode. Tower water flows through the jacket, and the steam is blocked. The tower water return is from the jacket exit, which is the highest temperature in the recirculation loop. The tower water enters at the recirculation pump suction, which is the lowest pressure within the recirculation loop.

Heating mode. Steam is admitted to the steam heater within the recirculation system to heat the water recirculating in the jacket. The tower water is blocked in the sense that there is no flow from tower water supply to tower water return. However, water continues to flow around the recirculation loop.

For the jacket configuration in Figure 3.20, the cooling mode uses tower water. The jacket configuration and the logic for transitioning between steam heating and one cooling mode (tower water, chilled water, or glycol) are largely the same. Only tower water cooling will be presented herein.

3.14.3. Control Configuration

For jackets with a recirculating loop, good control of vessel temperature is normally achieved using the cascade control configuration illustrated in Figure 3.21. This cascade configuration involves the following two controllers:

Jacket temperature controller (inner or slave controller). This temperature is usually sensed at the jacket outlet (the highest temperature in the

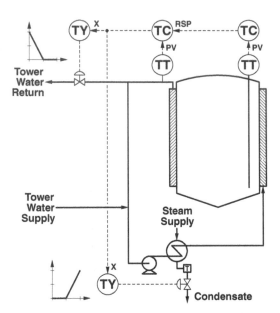

Figure 3.21. Split-range control configuration.

recirculation loop). Technically, this is the tower water return temperature. But with a small temperature rise from jacket inlet to jacket outlet,
the sensed temperature can be referred to as the jacket temperature.

Reactor temperature controller (outer or master controller). The controller output is the set point for the jacket temperature controller.

For the cascade configuration in Figure 3.21, another issue must be addressed
for the jacket temperature controller. This controller outputs to two final
control elements:

1. When cooling is required, the controller must output to a final control
   element associated with the cooling media. In Figure 3.21, this final
   control element is a control valve installed on the tower water return.
   During cooling, the heating media must be blocked.
2. When heating is required, the controller must output to a final control
   element associated with the heating media. In Figure 3.21, this final
   control element is a control valve installed on the condensate return.
   During heating, the cooling media must be blocked.

Other alternatives for the steam heater will be examined shortly.

At every instant of time, one media is blocked and the jacket temperature
controller is changing the opening of the control valve associated with the
other media. In terms of the flows, the transition from full cooling (steam

blocked; maximum tower water flow) to full heating (tower water flow blocked; maximum steam flow) is as follows:

1. Initially reduce the tower water flow; the steam remains blocked.
2. Once the tower water flow is zero (tower water is blocked), begin increasing the steam flow.

Full heating is achieved at maximum steam flow and tower water blocked.

3.14.4. Split-Range Control Logic

The control configuration in Figure 3.21 incorporates a methodology known as split range. The output of the jacket temperature controller is "split" at midrange. For the relationships illustrated in Figure 3.21, cooling occurs when the controller output is below midrange; heating occurs when the controller output is above midrange. The control valves behave as follows:

Controller Output	Cooling	Heating	Tower Water Control Valve	Condensate Control Valve
0%	Full	Off	Fully open	Completely closed
0–50%	Reducing	Off	Opening decreases	Completely closed
50%	Off	Off	Completely closed	Completely closed
50–100%	Off	Increasing	Completely closed	Opening increases
100%	Off	Full	Completely closed	Fully open

As the jacket temperature controller output changes from 0% to 100%, the jacket transitions from full cooling to full heating.

Figure 3.22 presents the valve openings as a function of the output of the jacket temperature controller. The representation in Figure 3.22 is for an ideal split-range implementation, where the transition from cooling to heating occurs at a controller output of exactly 50%. Practical considerations will be discussed shortly.

For the split-range logic configured as in Figure 3.22, the effect of changes in the controller output are as follows:

Below midrange. Increasing the controller output reduces the cooling, which will raise the jacket temperature.

Above midrange. Increasing the controller output increases the heating, which will raise the jacket temperature.

In both cases, the jacket temperature controller must be reverse acting—on an increase in the jacket temperature, the controller must reduce its output.

It is necessary that the directionality be the same for both heating and cooling. However, it is possible to swap the split-range logic so that heating

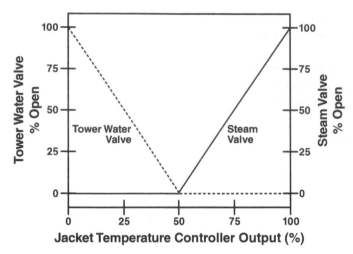

Figure 3.22. Ideal split-range logic.

occurs below midrange and cooling occurs above midrange. However, the only consequence is that the controller action must be changed.

3.14.5. Practical Considerations

Practical implementations deviate from the ideal. Factors such as errors in range adjustments mean that the actual behavior in the vicinity of a controller output of 50% is one of the following:

A deadband or gap in the control action. The tower water control valve closes slightly below 50%, and the steam control valve begins to open slightly above 50%. Within the deadband or gap, changes in the jacket temperature controller output have no effect on the jacket temperature. On a transition from cooling to heating and on a transition from heating to cooling, the consequence is an excursion in the jacket temperature from its target. The larger the gap, the greater the excursion.

An overlap in the control action. The tower water control valve closes slightly above 50%, and the steam control valve begins to open slightly below 50%. At 50%, both valves are slightly open. Cooling water is being admitted to the jacket only to be heated by steam. Although not especially efficient, the flows in the overlap region are small, so little is lost. The advantage is that changes in the jacket temperature controller output always affect the jacket temperature, giving a smooth and excursion-free transition between heating and cooling.

In some applications, the consequences of an overlap in the control action are of serious concern. To avoid an overlap, the range adjustments are intention-

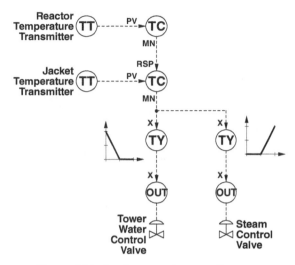

Figure 3.23. Implementation of split range.

ally biased so that a deadband or gap in the control action is assured. The issues are discussed in an accompany book [1] and will not be repeated here.

3.14.6. Implementing Split Range

With modern digital controls, split range is implemented as per the blocks illustrated in Figure 3.23 (or an equivalent structure). A separate output from the controls is provided for each control valve.

The advantages of implementing the split-range logic in software are as follows:

1. The split-range logic is expressed in terms of valve openings, which divorces the split-range logic from the fail-closed (air-to-open) versus fail-open (air-to-close) nature of the control valves.
2. The practical considerations pertaining to split range can be easily incorporated. Parameters such as deadbands or gaps can be easily adjusted, permitting the details of the split-range logic to be "fine-tuned" to the characteristics of the final control elements.

For each output, Figure 3.23 provides a characterization function or function generator (the "**TY**" elements) to translate the controller output to a valve opening. For the ideal split-range logic in Figure 3.22, the relationship of valve opening to controller output is as follows:

Tower water valve opening. A controller output of 0–50% is translated to a tower water control valve opening of 100–0%. For controller outputs above 50%, the tower water control valve opening is 0%.

Steam valve opening. A controller output of 50–100% is translated to a steam control valve opening of 0–100%. For controller outputs below 50%, the steam control valve opening is 0%.

For simple relationships as in the ideal split-range logic in Figure 3.22, other approaches could be used. However, characterization functions easily accommodate the practical issues that must be incorporated into the relationships.

The output of the split-range logic is the valve opening (percent open) for each valve. If the control valve is fail-open (air-to-close), the percent open value from the split-range logic must be converted to percent closed. There are two possibilities:

Valve block (or its equivalent). Element "**OUT**" in Figure 3.23 is a "valve block" that generates the actual output to the valve. The input to the valve block is always percent open. If the valve is fail-open (air-to-close), the valve block converts the valve block input to the percent closed value that is output to the valve.

Smart valve. When the output is transmitted to a smart valve via a communications network, the conversion of percent open to percent closed can occur in the smart valve.

The long-term solution will be to do the conversion in the smart valve. However, if the output is via a 4–20 mA current loop, 4 mA must correspond to valve fully open. Consequently, the valve block or its equivalent must be used, even if the field equipment is a smart valve.

3.14.7. Exchanger Configurations

Figure 3.24 illustrates three possible configurations for the steam heater. In selecting the appropriate configuration, a major requirement is that the

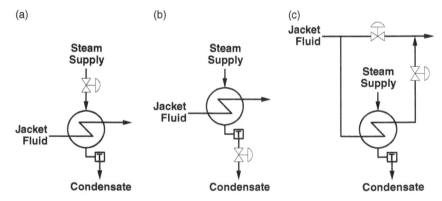

Figure 3.24. Control options for steam heater: (a) valve on steam supply; (b) valve on condensate; and (c) exchanger bypass.

transition from heating to cooling (and vice versa) must be smooth or bumpless. To achieve this, the steam heater must be capable of delivering very small heat transfer rates. Said another way, the minimum sustainable heat transfer rate must be essentially zero.

For each configuration, the water in the recirculation loop is being heated with steam. One issue that will be discussed shortly is cavitation within the recirculation pump. Consequently, the preference is usually the highest possible pressure within the recirculation loop. In Figure 3.21, the tower water control valve is located on the tower water return. This would be the case for all exchanger configurations in Figure 3.24.

3.14.8. Control Valve on Steam Supply versus Condensate

Consider the application described at the beginning of this section, specifically, a reacting system where a transition from cooling to heating is required late in the batch. The heat generated by the reaction is slowly decreasing. To maintain the reacting medium at the desired temperature, the cooling is slowly decreased until none is required, and then the heating must be slowly increased.

As explained in a previous section of this chapter, the exchanger in Figure 3.24a with the valve on the steam supply is not capable of low heat transfer rates when the temperature of the reacting medium is less than 100°C. Instead, there is a minimum heat transfer rate that can be sustained, that is, without cycling.

If the exchanger in Figure 3.24a is used, the transition from cooling to heating is as follows:

1. The cooling is slowly decreased until none is required. The split-range logic then starts opening the valve on the steam supply. When this transition occurs, the temperature throughout the recirculation system will be approximately equal to the temperature of the reacting medium.

2. Initially, the steam side of the exchanger is filled with condensate. No heat transfer occurs until the valve on the steam supply has opened sufficiently to force the condensate out of the exchanger. If condensate flows to a drain, a pressure slightly above atmospheric is required. If the condensate is returned to the boiler house, a somewhat higher pressure is required.

3. Once a sufficient pressure is attained to force the condensate from the exchanger, the steam side temperature is 100°C or slightly higher. The ΔT for heat transfer is this temperature less the temperature of the reacting medium. The result is the minimum heat transfer rate that can be sustained.

4. The heat transfer rate increases the temperature within the recirculation loop, which in turn increases the temperature of the reacting medium. The jacket temperature controller decreases its output, calling for less

heating or more cooling. The steam valve on the steam supply first closes, and then the control valve on the tower water supply opens to remove heat and lower the temperatures.

5. The temperatures drop below their desired values, and the cycle commences from step 1.

The net effect is that rate of heat removal is slowly reduced to zero, but then the rate of heat addition changes rather quickly to its minimum value. The jacket temperature control performance is as follows:

1. Smooth control of the jacket temperature is maintained until the switch to heating is required.
2. Cycling (alternately heating and then cooling) commences and continues until the required heat transfer rate is above the minimum heat transfer rate that can be sustained.
3. Smooth control is again achieved.

Locating the control valve on the condensate return as in Figure 3.24b enables the exchanger to sustain heat transfer rates very close to zero. Opening the condensate valve drains condensate from the exchanger, increasing the heat transfer surface area exposed to condensing steam. In the heat transfer equations, the ΔT term is constant (saturation temperature at the steam supply pressure less the temperature in the recirculation loop), but the area term increases as condensate is drained from the exchanger. Very low heat transfer rates are achieved by allowing the exchanger to be almost completely filled with condensate.

As compared with the configuration in Figure 3.24a with the control valve on the steam supply, the configuration in Figure 3.24b responds more slowly. Opening the condensate control valve has no immediate effect on the heat transfer rate. Instead, the heat transfer rate increases as condensate is drained from the exchanger. But for applications on jacketed vessels, the slower dynamics of the exchanger have little effect on the overall performance of the control loops. The overall dynamics are dominated by the thermal inertia of the reacting media and, to a lesser extent, the mass of water in the recirculation loop.

3.14.9. Exchanger Bypass

The configuration in Figure 3.24c provides a bypass for the steam-heated exchanger. The two control valves must be manipulated so as to not restrict the recirculation flow unnecessarily. To proceed from no heat to full heat, the approach is to first increase the opening of the exchanger valve and then decrease the opening of the bypass valve. This is reflected in the following split-range logic:

	Bypass Valve Opening	Exchanger Valve Opening
No heat	100%	0%
Increasing heat	100%	0–100%
Intermediate	100%	100%
Increasing heat	100–0%	100%
Full heat	0%	100%

Properly sizing the two control valves in Figure 3.24c is crucial. Improperly sized control valves can lead to regions where changing the opening of one of the valves has little effect on the heat transfer rate, basically resulting in a "dead zone" that degrades the performance of the controls.

On full cooling, the water flow through the steam heater is zero. Consequently, the water within the exchanger will be heated to the saturation temperature of the steam, with potentially the following adverse consequences:

The water must not vaporize. To prevent vaporization with no flow through the exchanger, the pressure in the recirculation system must exceed the steam pressure.

Minerals must not deposit on the heat transfer surfaces. The conditioning of the cooling water determines the maximum allowable temperature. Above that temperature, minerals will deposit.

Similar considerations apply to the recirculation system, which will be considered next.

3.14.10. Maximum Recirculation Water Temperature

For all of the exchanger arrangements in Figure 3.24, high temperatures in the recirculation loop could lead to recirculation pump cavitation. This could occur during both cooling and heating:

Cooling. For very low cooling water flows, the water in the recirculation loop would be heated to the reactor temperature.

Heating. The water in the recirculation loop could be heated to the saturation temperature at the steam supply pressure.

The maximum allowable temperature in the recirculation is the lower of the following:

Temperature at which scale forms on the heat transfer surfaces. This temperature depends on the conditioning of the cooling water. This value changes infrequently, if at all.

Temperature at which cavitation occurs in the recirculation pump. This is determined as follows:

1. The maximum permitted vapor pressure of the recirculating water is the pressure in the recirculation loop less the net positive suction head required by the pump. With the control valve on the tower water return, the pressure in the recirculation loop is the tower water supply pressure.
2. The temperature at which cavitation occurs is the temperature at which water exerts this vapor pressure.

Ideally, the tower water supply pressure is constant, but this is not assured in batch facilities where variations in the demand for cooling water routinely occur. When several units simultaneously require a high tower water flow, a decrease in the tower water supply pressure is a possibility, with the potential for cavitation in the recirculation pump.

The jacket temperature dynamics are relatively fast and can certainly track slow changes in tower water supply pressure. But very rapid drops in the tower water supply pressure are possible. To reduce the likelihood of cavitation during such events, an operating margin or tolerance can be provided. Specifically, the temperature computed to provide the required net positive suction head is reduced by some amount (a tolerance) to provide an operating margin, the objective being to avoid cavitation during rapid drops in tower water supply pressure. The value for the tolerance is basically a tuning parameter that is adjusted based on operational experience. The higher the tolerance, the less likely is pump cavitation. But setting the tolerance too high lowers the maximum heat transfer capability for the jacket.

3.15. SMOOTH TRANSITIONS BETWEEN TWO COOLING MODES

The vessel illustrated in Figure 3.25 has two cooling modes, tower water cooling and glycol cooling. The primary objective is to make a smooth transition between tower water cooling and glycol cooling. But in addition to the transition, the following issues must also be addressed:

1. Preferably remove as much heat as possible to the tower water, thus minimizing the use of glycol.
2. The tower water exchanger can potentially add heat to the recirculating glycol.
3. Water in the tower water exchanger freezes at the glycol supply temperature.

The first two are basically efficiency issues; the third results in physical damage to plant equipment.

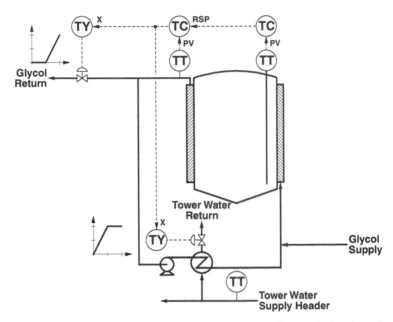

Figure 3.25. Jacket arrangement for tower water cooling and glycol cooling.

3.15.1. Split-Range Logic

The control configuration in Figure 3.25 relies on the following split-range logic to make a smooth transition between tower water cooling and glycol cooling:

	Controller Output	Tower Water Valve Opening	Glycol Valve Opening
No cooling	0%	0%	0%
Increasing cooling	0–50%	0–100%	0%
Max tower water cooling	50%	100%	0%
Increasing cooling	50–100%	100%	0–100%
Full cooling	100%	100%	100%

As increased cooling is required, the split-range logic first increases tower water cooling. When the maximum tower water cooling is attained, the split-range logic then increases the glycol cooling.

As compared with switching between steam heating and tower water cooling, the following differences arise:

1. For the vessel in Figure 3.21, the transition is made by reducing the tower water cooling to zero and then increasing the steam heating. Except for possibly a small overlap around midrange, one utility is blocked when the other is in use.

2. For the vessel in Figure 3.25, the transition is made by increasing the tower water cooling to its maximum value and then increasing the glycol cooling.

In the latter respect, the configuration in Figure 3.25 has an advantage of the switched jacket illustrated in Figure 3.19. The switched jacket is either cooling with tower water or cooling with glycol. Under most conditions, the configuration in Figure 3.25 removes as much heat as possible using tower water, leaving only the remainder (if any) to be removed using glycol.

3.15.2. Getting the Most from the Available Glycol

The incentives to optimize the use of glycol are as follows:

Glycol is a more expensive coolant than tower water. For specialty batch applications, utility costs are usually a small percentage of the production costs. If glycol cooling is required to manufacture the product, then it must be used. However, one does not want to incur extra costs unnecessarily.

The capacity of the refrigerant plant may be inadequate. In many plants, the peak demand for glycol can exceed the cooling capacity of the refrigeration plant. This is a consequence of the following:

1. The refrigerant plant capacity is usually sized based on the anticipated peak loads. Some assumptions are made regarding batch scheduling, but given long enough, times arise when these are not valid.
2. Plant enhancements to increase production capacity are not always accompanied by an increase in the refrigerant plant capacity.

In most batch facilities, increased plant productivity is the major incentive to use glycol cooling in a prudent manner. Utility costs are of less concern than a loss in production.

3.15.3. Heat Addition By Tower Water Exchanger

Under certain conditions, the tower water exchanger can be adding heat instead of removing heat. This depends on the following two temperatures:

Jacket temperature, or actually the jacket exit temperature. The recirculating glycol enters the tower water exchanger at this temperature.

Tower water supply temperature. A temperature transmitter in Figure 3.25 is installed on the tower water supply header. This temperature transmitter must be installed at a location where the flow is never stopped.

When the jacket temperature is lower than tower water supply temperature, the tower water exchanger would add heat to the recirculation loop. This heat has to be removed by increasing the glycol flow, which is not a prudent use of glycol.

As the jacket temperature approaches the tower water supply temperature, the heat removed by the tower exchanger approaches zero. Control logic must be added to block the tower water flow should the jacket temperature drop below the tower water supply temperature.

One possibility is to incorporate the following logic into the control configuration:

1. Compute ΔT_{IN} as the difference between the jacket temperature and the tower water supply temperature.
2. If $\Delta T_{IN} < 0$, block the tower water flow to the tower water exchanger.

Although relatively easy to incorporate into the control configuration in Figure 3.25, the following issues arise:

1. Abruptly blocking the tower water flow could cause a "bump" in the jacket temperature. In production reactors, the reactor temperature responds far more slowly, so the "bump" in the reactor temperature will be much less and probably of no consequence. Even so, "bumps" can become distractions and are preferably avoided.
2. Avoiding "chatter" (switching the tower water flow on and off on a short time interval) is essential. There are two options:
    (a) A deadband on the temperature difference that initiates the switch between no flow and full flow.
    (b) Minimum time between switches. For example, once the tower water flow is blocked, it must remain blocked for a specified time.

Another alternative is to gradually reduce the tower water flow as the jacket temperature approaches the cooling water supply temperature. This logic can be implemented as follows:

1. Compute ΔT_{IN} as the difference between the jacket temperature and the tower water supply temperature.
2. Based on ΔT_{IN}, the output of the characterization function is as follows:

Interval	Output
$\Delta T_{IN} < 0$	0%
$0 < \Delta T_{IN} < 2°C$	0–100%
$\Delta T_{IN} < 0$	100%

3. Use a low select to select between the following:
    (a) Output of the split-range logic that determines tower water cooling
    (b) Output of the previously mentioned characterization function.

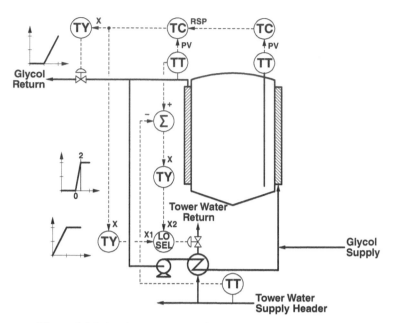

Figure 3.26. Logic to prevent adding heat from tower water.

Incorporating these into the split-range configuration in Figure 3.25 gives the configuration in Figure 3.26.

3.15.4. Freezing in Tower Water Exchanger

With the configuration in either Figure 3.25 or Figure 3.26, freezing in the tower water exchanger is a possibility. Potentially, glycol can lower the temperature within the recirculation loop below 0°C. Should this occur, the water in the tower water exchanger freezes, likely resulting in damage to the exchanger.

Should the glycol recirculation temperature drop to 0°C, actions must be initiated to avoid freezing. Two possibilities are the following:

1. Drain the tower water from the exchanger. This requires block valves on the tower water supply and return, plus piping and valves to drain water from the exchanger.
2. Flow sufficient tower water through the exchanger to prevent freezing. An override configuration could be configured so that the tower water control valve opening is adjusted so as to keep the tower water return temperature above freezing. A disadvantage of this approach is that the tower water exchanger will be adding heat to the glycol recirculation loop.

A simple alternative to the latter is to impose a specified opening on the tower water control valve should the glycol recirculation temperature drop below

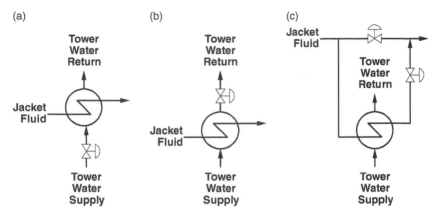

Figure 3.27. Control options for tower water exchanger: (a) valve on supply; (b) valve on return; and (c) exchanger bypass.

0°C. Configuring an alarm on the tower water return temperature would alert the process operators that a larger tower water flow is required. This is a "keep it simple stupid" approach, but is only acceptable when such conditions arise infrequently. This is possibly the case for a continuous process, but not for a batch process. For certain products, freezing conditions would arise for every product batch, making approaches that require no action from the process operators more desirable.

3.15.5. Alternate Exchanger Configurations

Figure 3.27 illustrates three possible configurations for the tower water exchanger (these are analogous to the configurations in Figure 3.24 for the steam heater). From a control perspective, configurations in Figure 3.27a,b are equivalent, the only difference being the pressure on the water side of the exchanger. As compared with these two, the bypass configuration in Figure 3.27c offers some advantages.

The two control valves in Figure 3.27c must be manipulated so as to not restrict the recirculation flow unnecessarily. To proceed from no cooling to full tower water cooling, the approach is to first increase the opening of the exchanger valve and then decrease the opening of the bypass valve. This is reflected in the following split-range logic:

Tower Water Cooling	Bypass Valve Opening	Exchanger Valve Opening
None	100%	0%
Increasing	100%	0–100%
Intermediate	100%	100%
Increasing	100–0%	100%
Full	0%	100%

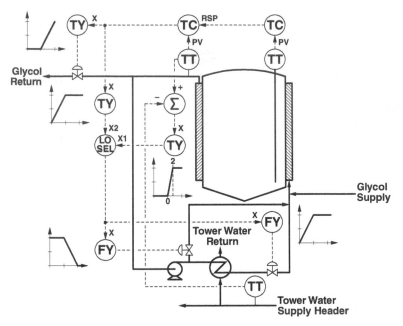

Figure 3.28. Split-range control configuration for exchanger with bypass.

3.15.6. Reactor Temperature Control

Figure 3.28 illustrates the reactor temperature control configuration when the bypass configuration is used for the tower water exchanger. The configuration consists of the following:

Split-range logic for jacket temperature controller. Below midrange, only tower water cooling is provided. Above midrange, the full tower water cooling is provided, supplemented by glycol cooling. The outputs of the two characterization functions are as follows:

1. Characterization function for tower water cooling outputs to the logic to prevent the tower water exchanger from adding heat, which in turn outputs to the split-range logic for the two valves associated with the exchanger.

2. Characterization function for glycol cooling outputs directly to the control valve on the glycol return (same as in Figure 3.25 and Figure 3.26).

Logic to prevent tower water exchanger from adding heat. Above midrange on the jacket temperature controller output, the tower water exchanger is on full cooling. But should the jacket temperature drop below the tower water supply temperature, the tower water exchanger would be heating the recirculating glycol. Under these conditions, the tower water exchanger should be switched to no cooling. The comparator, the characterization function, and the low selector in Figure 3.28 are the same as in Figure 3.26.

Split range for tower water exchanger bypass. The output from the low select is the input to this split-range logic. At 0% (no heating), the bypass valve is fully open, and the valve in series with the exchanger is fully closed. Between 0% and 50%, the exchanger valve opens, with the bypass valve also fully open. Between 50% and 100%, the bypass valve closes, with the exchanger valve fully open. At 100%, the bypass valve is fully closed, so all recirculating glycol passes through the exchanger to give the maximum heat transfer rate.

With the bypass arrangement, freezing in the tower water exchanger is not an issue. The behavior is as follows:

1. If the temperature of the recirculating glycol is less than the cooling water supply temperature, the logic to prevent the tower water exchanger from adding heat forces the input to the bypass split-range logic to be 0%. This condition will be imposed well before the temperature of the recirculating glycol approaches 0°C.
2. An input of 0% to the bypass split-range logic results in no recirculating glycol passing through the exchanger. With the tower water flowing rapidly through the exchanger, there is no risk of freezing.

3.15.7. Issues Pertaining to Bypass

If the objective is to remove as much heat as possible with the tower water, oversizing as follows should be considered:

1. Oversize the heat transfer surface area for the tower water exchanger.
2. Oversize the ability to flow water through the tower water exchanger.

There are two consequences:

1. The tower water temperature rise (supply to return) is very small.
2. Especially at heat duties below the maximum, the temperature of the glycol exiting the tower water exchanger is very close to the tower water supply temperature.

The tower cooling water is not being efficiently used, but the overriding consideration is to minimize the use of glycol.

As with all bypass configurations, properly sizing the valve in series with the exchanger and the valve in the bypass is crucial. The bypass split-range logic increases the heat transfer rate by first opening the exchanger valve and then closing the bypass valve. Below midrange on the jacket temperature controller output, the input to the bypass split-range logic is the output of the controller. From a control perspective, the preference is that the heat removal rate be linearly related to the input to the bypass split-range logic. In most cases, this relationship is sufficiently linear if the two valves are sized such that the fraction of the flow that passes through the exchanger is linearly related to the input to the bypass split-range logic.

REFERENCES

1. Smith, C. L., *Advanced Process Control: Beyond Single Loop Control*, John Wiley and Sons, New York, 2010.
2. Smith, C. L., *Practical Process Control: Tuning and Troubleshooting*, John Wiley and Sons, New York, 2009.

4

DISCRETE DEVICES

A major difference between continuous and batch plants is the ratio of discrete I/O points to analog I/O points. In continuous control applications, rarely does the number of discrete I/O points exceed the number of analog I/O points. But for most batch applications, the number of discrete I/O points exceeds the number of analog I/O points, possibly by a factor of 10.

In presentations pertaining to continuous processes, discrete I/O is rarely a major issue, often receiving barely a mention and sometimes no mention at all. For batch applications, discrete I/O and associated logic is of paramount importance. Consequently, this entire chapter is devoted to various issues pertaining to discrete I/O and discrete logic.

4.1. DISCRETE INPUTS

The value or state of a discrete input is 0 or 1, false or true, off or on, and so on. The signal for the discrete input is the output of a discrete measurement of some type. On the simple end, the discrete measurement may be a mechanical contact, a proximity switch, a photocell, and so on. Discrete measurements designed for process applications are generally more sophisticated. Often referred to as "switches," these include level switches, pressure switches, centrifugal switches on motor shafts, and so on. Figure 4.1 presents a vessel with two level switches. The one designated **LL** is installed to detect low level; the one designated **LH** is installed to detect high level.

Control of Batch Processes, First Edition. Cecil L. Smith.
© 2014 John Wiley & Sons, Inc. Published 2014 by John Wiley & Sons, Inc.

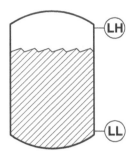

Figure 4.1. Level switches on a vessel.

Discrete measurements are said to be "actuated" or "not actuated." For most devices, not actuated is the same as the "shelf state"—if the device is removed from the process and placed on a shelf in the warehouse, the device is in its not actuated state. For devices that require power, the not actuated state is the condition when power is absent. However, this cannot be applied to all discrete measurement devices—level switches based on floats (such as in most toilet bowl tanks) do not require external power.

The discrete input from a process measurement reflects the state of the process. These process states can be indicated in two ways:

Binary value. These are 0/1, false/true, off/on, and so on. Herein 0/1 will be used. These values are the inputs to the control logic, but process operators find them inconvenient.

Names. These are meaningful to humans. The names can be any sequence of printable characters, preferably not so short as to be cryptic nor so long as to be verbose.

Using the level switches in Figure 4.1 as an example, the names for the states could be associated with the binary values for the states as follows:

Process State	Low-Level Switch	High-Level Switch
0	**Low Level**	**High Level**
1	**OK**	**OK**

The association of names with the binary values for the process states is entirely at the discretion of the user, so if desired, the opposite of the previously mentioned association could be used. However, there is one very important consideration: The selected association preferably applies to all devices of that type. That is, if state 1 means that the level indicated by a low-level switch is **OK**, then this statement should apply to all low-level switches in the plant. Associating **Low Level** with state 0 for some level switches but with state 1 for others leads to confusion (for humans, not the process controls).

4.1.1. Normally Open/Normally Closed

A push button can be the source of a discrete input. The push button is actuated by depressing the push button. If the push button is not depressed, the push button is not actuated. The state of the discrete input reflects the states of the push button—**Depressed** and **Not Depressed**.

A contact is incorporated in the push button. The discrete input senses the state of this contact, either **Open** or **Closed**. Push buttons are available in two styles:

Normally open. When the push button is not depressed, the contact within the push button is open. When the push button is on a shelf in the warehouse, the contact is open. Depressing the push button causes the contact to close.

Normally closed. When the push button is not depressed, the contact within the push button is closed. When the push button is on a shelf in the warehouse, the contact is closed. Depressing the push button causes the contact to open.

The style is specified at the time the push button is purchased.

The field wiring creates a circuit where the contact in the measurement device is essentially in series with the discrete input circuitry within the controls. The input hardware can determine whether or not continuity exists in this electrical circuit. The value or state of the hardware input is as follows:

Input state 0. Continuity is not present.
Input state 1. Continuity is present.

How does one decide which type of contact to purchase? Suppose a push button is used for an emergency stop. The options are as follows:

Normally open push button. Emergency stop is initiated on the presence of continuity in the circuit. Continuity exists when the push button is depressed, giving a hardware input state of 1. Otherwise, the hardware input state is 0.

Normally closed push button. Emergency stop is initiated on the absence of continuity in the circuit. Continuity does not exist when the push button is depressed, giving a hardware input state of 0. Otherwise, the hardware input state is 1.

The push button is depressed to initiate an emergency stop, which hopefully does not occur too often. However, the emergency stop must occur when the push button is depressed.

A "broken wire" is any fault in the field wiring whose consequence is that continuity can never be present in the circuit. The consequences are as follows:

Normally open push button. The fault will not be detected until the emergency stop is depressed. Continuity is required to initiate the emergency stop, but with the fault in the wiring, continuity cannot exist.

Normally closed push button. The fault is equivalent to depressing the emergency stop push button. The fault causes an emergency stop to occur immediately.

A normally closed push button must be installed. An emergency stop not occurring when the push button is depressed is unacceptable. Emergency stops occurring unnecessarily are undesirable, but wiring faults are infrequent (if not, there are ways to address this problem).

Approaches are available that can distinguish between an input state of 0 and a broken wire. The measurement device functions as follows:

Contact open. A large electrical resistance is inserted into the circuit.

Contact closed. A small electrical resistance is inserted into the circuit.

The following can be distinguished:

Broken wire. The current flow through the circuit is zero.

Hardware input state 0. A "small" current flows through the circuit.

Hardware input state 1. A "large" current flows through the circuit.

The distinction between small and large depends on the voltage imposed on the circuit by the input hardware and the resistances inserted into the circuit by the discrete measurement.

Such approaches are widely used in residential low-voltage alarm systems but not in industrial applications. Consider the case of the emergency stop push button described previously. Does being able to distinguish between broken wire and contact open offer any advantages? Consider the two options for the push button:

Normally closed. By distinguishing between broken wire and push button depressed, the option is available for not initiating an emergency stop on a broken wire. But with the broken wire condition in effect, depressing the push button cannot be detected, which means that the emergency stop does not function. Rarely could a facility be operated without a functioning emergency stop.

Normally open. The broken wire condition can be detected when the push button is not depressed, but what action must be taken? An alert could be generated immediately to the effect that the emergency stop is not functional. But rarely, if ever, could operations be continued without a functioning emergency stop.

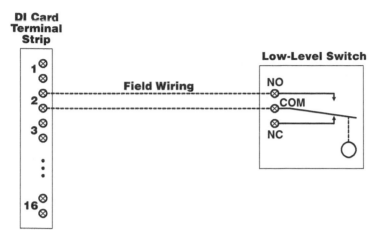

Figure 4.2. Wiring for the low-level switch.

In both cases, the process must be stopped on the occurrence of a broken wire. Provided the push button is normally closed, the ability to distinguish a broken wire from contact open offers no advantages.

4.1.2. Process Switches

More sophisticated discrete measurements, such as the switches designed for process applications, provide three terminals for connecting the field wiring. Two contacts are provided in the "double pole, single throw" configuration illustrated in Figure 4.2. The terminals are customarily designated as follows:

- Normally closed (NC)
- Common (COM)
- Normally open (NO).

When the discrete measurement is in its "normal" state, the following statements apply:

- Electrical continuity is present between the normally closed terminal and the common terminal—the "normally closed contact" is said to be closed.
- Electrical continuity is not present between the normally open terminal and the common terminal—the "normally open contact" is said to be open.

The field wiring connects one of these "contacts" with a hardware discrete input point whose type is "contact sense." Figure 4.2 illustrates the normally open contact of the low-level switch wired to point 2 on the discrete input terminal strip. For the low-level switch, a low-level condition within the process gives the following results:

Wired to the normally open contact (as in Figure 4.2). Low level results in a hardware input state of 0 (continuity not present).

Wired to the normally closed contact. Low level results in a hardware input state of 1 (continuity present).

From the perspective of the application software, either approach is acceptable. But as will be explained shortly, failure considerations favor wiring to the normally open contact of the low-level switch.

4.1.3. Sense Mode

The following distinction is important:

* Process state as used in the control logic and displayed to the process operator
* Hardware discrete input state as sensed from the discrete measurement.

State 1 for the hardware input from a discrete measurement does not necessarily mean that the process state is also state 1. This association depends on the type of contact (normally open or normally closed) that provides the hardware input.

For the low-level switch in Figure 4.2, the significance of the discrete input depends on the terminal to which the wiring is connected:

Contact for Wiring	Input State 0	Input State 1
Normally open	**Low Level**	**OK**
Normally closed	**OK**	**Low Level**

For the high-level switch, the significance is as follows:

Contact for Wiring	Input State 0	Input State 1
Normally open	**OK**	**High Level**
Normally closed	**High Level**	**OK**

Using low-level switch as an example, one could propose the following approaches:

1. Assign the names that are to apply to the process states indicated by all low-level switches as follows:
   Process state 0: **Low Level**
   Process state 1: **OK**.

2. For every low-level switch in the plant, connect the field wiring to the appropriate contact so that a low level in the process results in state 0 for the hardware input.

For reasons to be explained shortly, this may not give the most desirable wiring arrangement for some of the low-level switches. That is, for some low-level switches, the preferable wiring is for a low-level condition in the process to give a discrete input state of 1.

With modern digital control systems, the same process state can be obtained with the wiring connected to either terminal. The association of process states with hardware input states is determined by a parameter referred to as the sense mode. Herein the sense mode will be designated as either "direct" or "reverse," the significance being as follows:

Direct. Process state is the same as the hardware input state. That is, a hardware input state of 0 gives a process state of 0.

Reverse. Process state is the opposite (logical NOT) of the hardware input state. That is, a hardware input state of 0 gives a process state of 1.

The purpose of the sense mode is to decouple the following two activities:

1. The process people can associate the process states (0 or 1) with conditions within the process (as reflected by the names for the states) in whatever manner they choose.
2. The input can be wired in the most appropriate manner.

On low level, the low-level switch in Figure 4.1 is not actuated. The following give the same result for the process state:

Contact for Field Wiring	Hardware Input State for Low Level	Sense Mode	Process State
Normally open	0	Direct	0—**Low Level**
Normally closed	1	Reverse	0—**Low Level**

On high level, the high-level switch is actuated. For the high-level switch, the following give the same result for the process state:

Contact for Field Wiring	Hardware Input State for High Level	Sense Mode	Process State
Normally open	1	Reverse	0—**High Level**
Normally closed	0	Direct	0—**High Level**

4.1.4. Process Normal

For the normally open contact and the normally closed contact, the concept of normal should be considered as "equipment normal" and must not be confused with "process normal." For the vessel in Figure 4.1, process normal means that the liquid level is above the low-level switch and below the high-level switch. Level switches are actuated when liquid is present at the location of the level switch. When conditions within the process are normal, the states of the level switches are as follows:

Low-level switch. This switch is actuated, and consequently is not in its equipment normal state. Continuity is present at the normally open contact; continuity is not present at the normally closed contact.

High-level switch. This switch is not actuated, and consequently is in its equipment normal state. Continuity is present at the normally closed contact; continuity is not present at the normally open contact.

When implementing the field wiring, the controlling principle is the following:

Continuity in the circuit must indicate the process normal condition.

For the level switches, the selection is as follows:

Low-level switch. Connect the wiring to the normally open contact. For process normal, the low-level switch is actuated, so continuity is present at the normally open contact.

High-level switch. Connect the wiring to the normally closed contact. For process normal, the high-level switch is not actuated, so continuity is present at the normally closed contact.

Why is wiring in this manner desirable? One possible failure is a fault in the field wiring, that is, a broken wire. This results in no continuity in the circuit and a hardware input state of 0. For the low-level switch in Figure 4.1, the consequences are as follows:

Wired to normally open contact. A hardware input of 1 indicates the **OK** state. A break in the field wiring results in a hardware input of 0, resulting in a **Low-Level** indication. The consequence is a false indication of a problem in the process. A **Low Level** is surely an alarm condition, and may initiate other actions (such as blocking discharge flows). Although the indicated problem is different from the real problem, at least something happens, thereby making people aware that a problem exists.

Wired to normally closed contact. A hardware input of 0 indicates the **OK** state. A break in the field wiring results in a hardware input of 0, so the **OK** indication continues. Should a low-level condition arise in the

process, the low-level switch responds, but the fault in the field wiring
causes the indication to be **OK** regardless of the state of the low-level
switch. The consequence is a problem in the process that is not reported
to the controls. So nothing happens! A problem exists, but people think
that all is well.

Especially in critical applications, the latter is not acceptable. That is, it is better
to respond to a false report of a problem than to not respond to a real problem.
Basically, the process condition is unknown, so assume the worst.

When conventional field wiring is replaced by a communications network,
the counterpart to a fault in the wiring is a failure in the communications with
the discrete measurement. A communications failure must result in the same
response as a fault in the wiring; specifically, on a loss of communications, the
process state must be reported as **Low Level**. Again, the process condition is
unknown, so assume the worst.

Especially when personnel safety is involved, adhering to the previous statement regarding wiring is mandatory. But when the only consequence is "making
a mess," equipment damage that can be quickly repaired, and so on, adhering
to this statement is desirable, but deviations are often tolerated. Systems pertaining to personnel safety are usually simple, and considerable effort is
expended to avoid errors. Controls for batch facilities have a far larger number
of inputs (often exceeding 1000), giving so many opportunities to make a
mistake that a few of them will be taken. Changing the software is far easier
than changing the hardware, especially during plant commissioning.

4.2. DISCRETE OUTPUTS

Controls for batch applications provide discrete outputs to drive various two-
state final control elements that largely consist of the following:

1. Two-state valves, often referred to as remote control valves (RCVs). The
   actuator can only position these valves at the extremes, that is, either
   fully open or fully closed. The actuator may be air operated or electrical
   operated (solenoid valves).
2. On–off motors. These are fixed-speed motors that are either running full
   speed or stopped.

Others are occasionally encountered. For example, a pneumatic cylinder can
be installed to open a door. The result is either a fully closed door or a door
that is opened to the extent permitted by the cylinder. However, the discussion
herein considers only valves and motors.

The states of final control elements are assigned binary values such as 0/1,
false/true, and so on. In turn, these are associated with names that are meaningful to humans, examples being the following:

State	Valves	Motors
0	**Closed**	**Stopped**
1	**Open**	**Running**

These are the states of the final control elements and are not necessarily the same as the state of the hardware output to that final control element.

4.2.1. Output Configurations

Figure 4.3 presents the following three options for the interface between the controls and two-state field device:

1. Single latched discrete output
2. Dual momentary discrete outputs
3. One latched output, one momentary output (sometimes referred to as the enable/operate configuration).

These three options also apply to valves, on–off motors, and so on.

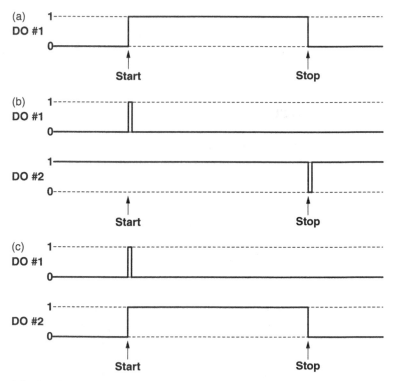

Figure 4.3. Configurations for discrete output hardware: (a) latched; (b) dual momentary; and (c) one latched, one momentary.

Why the different configurations? Digital process controls in industrial processes did not start from a blank page—hardwired circuits and panels were installed in all existing plants and were an option in new plants. Furthermore, people did not completely trust the new technology, and a few even thought it had absolutely no future in process applications (after all, they told Orville and they told Wilbur that man was not destined to fly through the air). This led to the requirement that the panels be retained and, if necessary, operations personnel be able to operate all field devices from the panels. Although panels have largely disappeared, certain aspects of their legacy remain. Once a certain approach is adopted, it tends to be continued unless specific reasons arise to justify making a change.

To appreciate a couple of considerations pertaining to the selection of the output configuration, a few simple wiring diagrams for operating field devices will be presented. With hardwired logic and panels, operators opened and closed valves, started and stopped motors, and so on, using push buttons. Push buttons can be purchased in a variety of styles and configurations. Figure 4.4 presents the representation used in wiring diagrams for four styles of push buttons:

Momentary. The push button is actuated only when depressed. As soon as the operator ceases to press the push button, the push button is not actuated. There are two options:

> **Normally closed.** Continuity is present when the push button is not depressed (its normal or not actuated state). Depressing the push button breaks the continuity, but only while the push button remains depressed.

> **Normally open.** Continuity is not present when the push button is not depressed (its normal or not actuated state). Depressing the push button provides the continuity, but only while the push button remains depressed.

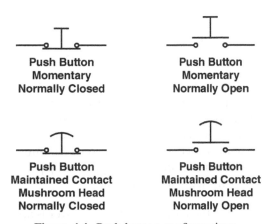

Figure 4.4. Push button configurations.

Maintained or latched. The push button is actuated when depressed. Once actuated, a mechanical latching mechanism maintains the actuated state when the push button is no longer depressed. Releasing the latch requires a specific action from the operator, such as depressing the push button again, rotating the button a quarter turn, or some other action that depends on the nature of the latching mechanism within the push button. The symbol in Figure 4.4 is for the "mushroom head" style of push button, which are commonly available with such a latching mechanism.

Normally closed. Continuity is present when the push button is not actuated (its normal or not actuated state). Depressing the push button breaks the continuity, and the continuity remains broken until the operator performs the action required to release the latching mechanism within the push button.

Normally open. Continuity is not present when the push button is not depressed (its normal or not actuated state). Depressing the push button provides continuity, and a mechanical mechanism within the push button maintains the continuity until the operator performs the action required to release the latching mechanism within the push button.

Alternate representations for push buttons are available. The ones in Figure 4.4 is an older style, but one that remains in common use.

4.2.2. Latched Configurations

Figure 4.5a illustrates the use of a single push button to operate a field device. The push button must be of the maintained or latched style. The circle labeled "field device" is an electrical coil that actuates when voltage is applied. For starting or stopping a motor, actuating the coil applies power to the motor. For opening or closing a valve, actuating the coil applies power to the actuator for the valve. Depressing the push button in Figure 4.5a opens the valve or starts the motor. The field device remains in this state until the latching mechanism within the push button is released.

The wiring diagram for many field devices must provide for an interlock, an emergency stop, and/or local operation, all of which will be discussed in detail in a later section of this chapter. The circuit in Figure 4.5b contains a normally closed maintained push button to provide an emergency stop. When not depressed, continuity is provided, and the field device can be operated. When depressed, continuity is broken, and no power can be applied to the output coil for the field device. Once depressed, continuity remains broken until the latching mechanism within the emergency stop push button is released.

Consider the following sequence of events:

1. The valve is open or the motor is running (power is applied to the output coil).

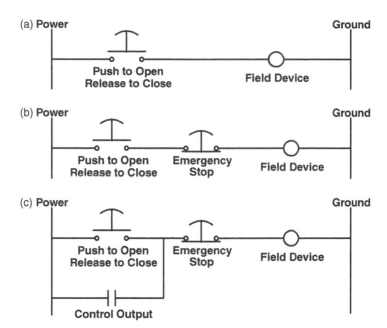

Figure 4.5. Latched push button arrangements: (a) latched push button; (b) latched push button and emergency stop; and (c) latched push button, latched control output, and emergency stop.

2. The emergency stop push button is depressed, causing the valve to close or the motor to stop.
3. At some time later, the latching mechanism within the emergency stop push button is released.

Now for the important question: how does the valve or motor respond? The mechanical latching mechanism within the push button to operate the field device is not automatically released when the emergency stop push button is depressed. If this push button is still latched when the emergency stop push button is released, the valve immediately opens or the motor immediately starts. This is not a desirable situation—valves should open, motors should start, and so on, only in response to either an operator action or an action by the process controls.

Suppose the requirement is that the field device can be operated either from the process controls or by the operator using the panel. Figure 4.5c wires the output contact from the process controls in parallel with the push button that operates the field device. However, the behavior of this circuit is not acceptable. The following statements are true:

1. If the operator's push button is not actuated, the process controls can operate the field device.

2. If the output contact from the process controls is not providing continuity, the operator can operate the field device.

However, the following are also true and are not acceptable:

1. If the operator has depressed the push button in the panel, the process controls cannot operate the field device. With continuity through the push button, the presence or absence of continuity through the output contact for the process controls has no effect on the state of the field device.
2. If the process controls is providing continuity through its output contact, the process operator cannot change the state of the field device. With continuity through the output contact for the process controls, the presence or absence of continuity through the push button in the panel has no effect on the state of the field device.

Such behavior is not acceptable—if the operator opened a valve or started the motor, the process controls must be able to close the valve or stop the motor, and vice versa.

To provide acceptable performance, a switch can be incorporated into the circuit in Figure 4.5c to provide two modes of operation:

Local. The operator operates the field device from the panel.
Remote. The process controls operate the field device.

This permits either to operate the device, but not both at the same time. In practice, there is some merit in this. However, this raises the issue of bumpless transfer between the local and remote modes. That is, the field device must not abruptly change states when the mode is switched from local to remote, or vice versa. Tracking can be incorporated into the process controls to achieve a smooth switch from local to remote, but the inability to effect tracking for the mechanical latch in a push button means that the switch from remote to local may not be bumpless.

Removing the requirement to support the panel eliminates the need in Figure 4.5c for the maintained push button to operate the field device. A single latched output from the process controls can be used to operate the field device. A discrete input is required to indicate the state of the emergency stop push button, and is usually justified so that the control room operators can be alerted that an emergency stop is in effect. This same input can be used to track the state of the latched outputs. For an emergency stop, the associated field devices assume their nonpowered or safe state. For these, tracking involves setting the latched discrete outputs to state 0, which means that these field devices will not change states when the emergency stop is released or reset. More on this later.

4.2.3. Momentary Configurations

To operate a field device, Figure 4.6a illustrates the use of the following two momentary push buttons:

1. Depressing the "Open" or "Start" push button opens the valve or starts the motor. This is a normally open push button.
2. Depressing the "Close" or "Stop" push button closes the valve or stops the motor. This is a normally closed push button.

Although the push buttons do not provide latching, the circuit in Figure 4.6a provides latching, the sequence being as follows:

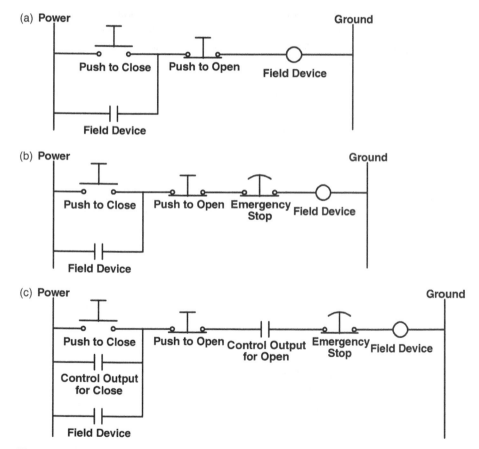

Figure 4.6. Momentary push button arrangements: (a) momentary push buttons; (b) momentary push buttons and emergency stop; and (c) momentary push buttons, momentary control outputs, and emergency stop.

1. Depressing the Open or Start push button causes the power to be applied to the output coil.
2. A contact associated with the output coil closes when power is applied. Power is being applied to the output coil through the push button and through the contact on the output coil.
3. After the Open or Start push button is no longer depressed, the output coil remains energized because power is being applied through the contact on the output coil.
4. Depressing the Close or Stop push button removes power from the output coil, which also causes the contact associated with the output coil to open.
5. After the Close or Stop push button is released, power is no longer applied to the output coil.

Such arrangements were commonly installed in conventional panels.

The circuit in Figure 4.6b contains a normally closed maintained push button to provide an emergency stop. Consider the same sequence of events as before:

1. The valve is open or the motor is running (power is applied to the output coil).
2. The emergency stop push button is depressed, causing the valve to close or the motor to stop.
3. At some time later, the latching mechanism within the emergency stop push button is released.

Now how does the valve or motor respond? Depressing the emergency stop push button removes power to the output coil, so the valve closes or the motor stops. This also opens the contact associated with the output coil, which releases the latch provided within the circuit. When the emergency stop push button is released, power is not applied to the output coil, and the field device does not abruptly change state. The same result is obtained with a momentary push button for the emergency stop. However, this would permit someone to immediately open the valve or start the motor. This is undesirable, so the emergency stop push button in circuits such as 4.6(b) are normally latched or maintained. The field device cannot be operated until the latching mechanism within the emergency stop push button is released.

The circuit in Figure 4.5c permits the field device to be operated by the operator and from the process controls. Two outputs are required from the process controls:

"Open" or "Start" output. Momentarily closing this contact applies power to the output coil, causing the valve to open or the motor to start. This contact is wired in parallel with the Open or Start push button. This contact should be open except when opening the valve or starting the motor.

"Close" or "Stop" output. Momentarily opening this contact removes power from the output coil, causing the valve to close or the motor to stop. This contact is wired in series with the Close or Stop push button. This contact should be closed except when closing the valve or stopping the motor. Some systems do this in software; others do it on the output card (the output point can be configured so that 0 means output contact closed and 1 means output contact open).

The circuit in Figure 4.5c permits either the process operator or the process controls to operate the field device at any time. This is not necessarily a good idea—the desire is for either to be able to operate the field device, but not at the same time. A switch can be incorporated into the circuit in Figure 4.5c to provide a **Local** and a **Remote** mode of operation. With momentary push buttons and momentary outputs from the process controls, the switch is inherently bumpless. The reasons are the same as for the emergency stop. But as panels disappeared, the requirement for **Local** and **Remote** modes of operation also disappeared, at least for most field devices.

4.2.4. Latched/Momentary Configurations

Batch operations continue to require activities that are largely manual. Consider transferring material from one or more drums to a process vessel. When multiple drums are required, transferring from one drum must be stopped, a full drum must be set up, and the transfer resumed from the full drum. Since this is largely a manual activity that can only be performed in the field, doing it entirely from the field has advantages. That includes stopping and subsequently starting the transfer pump, and possibly closing and subsequently opening one or more valves.

If the process controls have started the transfer, the operator should be able to stop it and then restart it. However, the operator should not be able to start the transfer unless the process controls previously started it. This capability is easily provided by a slight modification to how the Close or Stop contact in Figure 4.5c is used. Basically, logic for this contact is changed to latched instead of momentary, with the output states having the following significance:

State 0 (contact is open). The valve is closed or the motor is stopped. Furthermore, the valve cannot be opened or the motor started using the operator's Open or Start push button.

State 1 (contact is closed). The valve or motor can be opened using the operator's push buttons. If the valve is open or the motor is running, depressing the Close or Stop push button closes the valve or stops the motor. Provided this has no effect on the outputs from the process controls, the operator can subsequently use the Open or Start push button to open the valve or start the motor.

Basically, this output becomes a permissive for the valve to open or for the motor to start. Setting this output to state 0 will close the valve or stop the motor. However, setting this output to state 1 does not open the valve or start the motor. When used in this manner, the terms Close or Stop for this output are not entirely accurate.

To open the valve or start the motor, the process controls must do the following:

"Close" or "Stop" output. Set the state of this output to 1. This permits the valve to open or the motor to start. In a sense, this output is an "Enable."

"Open" or "Start" output. Momentarily set the state of this output to 1. This causes the valve to open or the motor to start. In a sense, this output is a "Drive" or "Operate."

To close the valve or stop the motor, the controls need only to set the state of the Close or Stop output to 0.

Using the outputs in this manner provides the desired behavior, but is not entirely free of issues. Consider the transfer pump. When the process controls start the pump, its state would be displayed as **Running**. But when the operator stops the pump via the circuit in Figure 4.5c, some issues arise.

To appropriately display the state of the transfer pump, an input indicating its state must be provided. As will be discussed shortly, this input is called a "state feedback" and is normally provided. Basically, the input is obtained via a contact on the output coil of the circuit in Figure 4.5c. To display the correct state of the field device, the display must be based on this input instead of the outputs from the process controls.

Also subsequently discussed is the "discrete device driver," one of its functions being to detect differences between the field device state as determined by the outputs from the process controls and the field device state as indicated by the state feedbacks. When the operator stops the transfer pump using the circuit in Figure 4.5c, a discrepancy is properly detected and an alert generated to the process operator. This alert is generally understood to be a problem within the final control element, which is not the case. The operator has stopped the transfer pump using the Stop push button provided for this purpose. This is an expected activity, making any alerts issued in response to this to be undesirable (they become "nuisance alarms").

When the emergency stop is depressed, the outputs from the process controls should track the states of the field devices. For the latched/momentary output configuration, the latched output should be set to state 0, which prohibits the field device from being started. But when the operator stops the transfer pump using the Stop push button, such tracking is not appropriate. If the process controls set the latched output to state 0, the operator cannot restart the transfer pump using the Start push button.

These considerations usually make another approach preferable. Instead of incorporating the Start and Stop push buttons into the circuit as in Figure 4.5c, the push buttons provide discrete inputs to the process controls, which then are configured to appropriately respond to the operator's requests to stop and start the transfer pump. With the process controls determining the state of the field device at all times, tracking is not required, and no discrepancy occurs between the state feedbacks and the control outputs that would cause the discrete device driver to generate an alert.

4.2.5. Role of Programmable Logic Controllers (PLCs)

In the early days of digital process controls, circuits such as those in Figure 4.4 and Figure 4.5 were hardwired, often within the same panels that contained the push buttons. But with the rapid acceptance of PLCs in industries such as automotive, those in the process industries discovered that it was more economical to implement such circuits in a PLC. The ease of troubleshooting provided even more incentive to migrate to PLCs.

These PLCs were low-end products whose role was limited to replacing hardwired circuits such as those in Figure 4.4 and Figure 4.5. Being simple devices, their reliability was accepted long before the reliability of the more complex process controls was accepted. Such PLCs also began to be used to implement safety interlocks, with most companies accepting this approach as an alternative to hardwired circuits. But often there was a caveat—the PLC had to be dedicated to such functions. That is, piggy-backing process controls onto such a PLC led to excessive complexity for a system in which safety interlocks were being implemented.

The introduction of such PLCs also led to changes in the approach to functions such as emergency stops. Instead of being incorporated into hardwired circuits, the emergency stop provided an input to the PLC which in turn provided all functions required for the emergency stop. Instead of a maintained push button for the emergency stop, two momentary push buttons were an alternative, one to impose the emergency stop and one to reset the emergency stop.

4.3. STATE FEEDBACKS

The logic for batch applications often requires a positive indication that a final control element is in a given state. The purposes include the following:

1. Before initiating and during certain activities within the plant, certain final control elements must be in their appropriate states. This will be illustrated shortly for valves on a piping header.

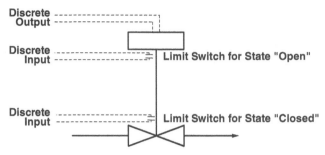

Figure 4.7. Two-state valve outfitted with two limit switches.

2. Advise operations personnel that a final control element may not be functioning properly.

The final control element is equipped with the required hardware to provide one or more discrete inputs that provide information pertaining to the state of the final control element. These signals are generically referred to as "state feedbacks."

4.3.1. Two-State Valves

Figure 4.7 illustrates a two-state valve equipped with two limit switches. Suppose the states are designated as follows:

State 0: **Closed**
State 1: **Open**.

One limit switch (the state 0 feedback) confirms that the valve is in the **Closed** state; the other limit switch (the state 1 feedback) confirms that the valve is in the **Open** state.

In older installations, the limit switches were physically mounted much as suggested by Figure 4.7. For a sliding-stem valve, a bar was attached to the valve stem. When the valve is in the **Closed** state, the bar physically closes the contact for the **Closed** state. When the valve is in the **Open** state, the bar physically closes the contact for the **Open** state.

Contacts mounted in this manner were subject to numerous problems, mainly because they were exposed to being physically struck in a variety of ways. The contacts could be damaged, or they could be repositioned slightly so that they would not close when the valve was in the corresponding state. Over the years, improvements such as replacing mechanical contacts by proximity switches enhanced their reliability, but problems remained to the extent that if a valve was commanded to close and the limit switch did not confirm that the valve was closed, the prime suspect was a limit switch.

This situation continued until the actuator manufacturers began incorporating the limit switches into the actuator housing. The compact nature and the reliability of proximity switches made this technically attractive. And customers were also demanding that the actuator manufacturers address the problem. With the limit switches in the actuator, the inputs actually confirm that the actuator is in the requested state, not that the valve itself is in the requested state. Failures such as a broken valve stem are possible but sufficiently rare that most users are comfortable with the limit switches physically located within the actuator housing.

While Figure 4.7 and this discussion pertain to sliding stem valves, comparable configurations are available for rotary stem valves.

The significance of the states of the discrete inputs from the limit switches is as follows:

State Confirmed by Limit Switch	Discrete Input Value Is State 0	Discrete Input Value Is State 1
State 0 (**Closed**)	Not fully **Closed**	**Closed**
State 1 (**Open**)	Not fully **Open**	**Open**

Does "Not fully **Closed**" mean the state of the valve is **Open**? Does "Not fully **Open**" mean the state of the valve is **Closed**? Probably, but not necessarily— the valve could be "stuck" between fully closed and fully open. In critical applications where one must be absolutely certain that the valve is in a given state, a limit switch must be installed to give a positive indication that the valve is in that state.

4.3.2. Final Control Element States

The availability of two limit switches permits the following assessment of the state of the final control element:

Limit Switch on State "Closed"	Limit Switch on State "Open"	State of Field Device
0	1	**Open**
1	0	**Closed**
0	0	**Transition**, then **Invalid**
1	1	**Invalid**

Starting with the valve closed, a command to open the valve should result in the following sequence of states:

State of Final Control Element	Limit Switch on State "Closed"	Limit Switch on State "Open"
Closed	1	0
Transition	0	0
Open	0	1

The **Transition** state should be a temporary state that is permitted to exist for a finite period of time called the transition time. The transition time is specified individually for each final control element and, for a valve, reflects the travel time of the valve. This could be as short as 10 seconds for a small solenoid valve, but 2 minutes or more for a large, motor-driven valve.

If the transition time elapses and the value of the input from each limit switch is 0, the state of the final control element changes to **Invalid**. The **Invalid** state is an abnormal state and the following ensue:

1. An alert is issued to the process operator.
2. This situation is considered to be a failure, so any control actions that rely on that final control element are appropriately suspended or terminated.

A transition time of 20 seconds does not mean that the valve remains in the transition state for 20 seconds following any change in its state. The transition state should be terminated as soon as the limit switches confirm that the valve is in the state corresponding to the current values of the output to the valve. While one should not specify excessively large values for the transition times, being conservative is the usual practice.

4.3.3. Two-State Motors

Figure 4.8 illustrates a two-state motor equipped with only one state feedback. This is a two-state field device, the states being designated as follows:

State 0: **Stopped**
State 1: **Running**.

Only one state feedback is provided and is normally considered to confirm that the motor is **Running**.

In reality the interface is to the motor starter circuitry, not to the motor itself. One or more discrete outputs cause the motor starter to either apply power to the motor or disconnect power from the motor. The indication provided by the state feedback is as follows:

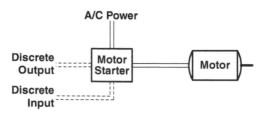

Figure 4.8. Two-state motor outfitted with one state feedback.

Value of discrete input is 0. Power is not being applied to the motor.

Value of discrete input is 1. Power is being applied to the motor. Although it is possible that the motor is not truly running, the occurrences of this are unusual.

Normally, this input is configured for state 1—field device is **Running**. A transition time of only a few seconds is required. This time reflects the time for the motor starter circuit to respond by applying power to the motor; it does not reflect the time required for the motor to attain full speed.

If it is necessary to confirm that the motor is running or not, a centrifugal switch must be installed on the motor shaft to indicate that the rotational speed is above or below a set value. This is most commonly installed when assurance is required that whatever is being driven by the motor is no longer rotating. Once power is disconnected from the motor, rotating devices tend to "spin down," with the time required to spin down depending on the inertia of the rotating equipment. In such cases, the logic first disconnects power from the motor, causing the state feedback from the motor starter circuit to indicate that the state of the motor is **Stopped**. However, this does not mean that the equipment is no longer rotating. If it is essential that the rotation has stopped before proceeding further, the logic must wait until the input from the centrifugal switch confirms that the equipment is no longer rotating.

4.3.4. Discrete Device Driver

Given the number of two-state final control elements in the typical batch plant, some thought must be devoted as to how to implement the logic. To support the state feedbacks, the following capabilities are required:

1. Support various configurations (no state feedbacks, state 0 feedback only, state 1 feedback only, and two state feedbacks).
2. Provide a transition time, preferably specified individually for each final control element.
3. Determine the final control element state as state 0, state 1, transition, or invalid.
4. Support various output configurations (latched, dual momentary, enable/operate, etc.).

In practice, this is the minimum; additional desirable capabilities will be discussed shortly.

The options for implementing the required logic are as follows:

Individually program the logic for each final control element. The early versions of PLCs did not support subroutines (or their equivalent), so specific logic had to be provided for each final control element. The PLC programmers were normally instructed to follow a "standard" provided for each feedback configuration. An advantage of this approach is that troubleshooting could be performed in the manner typical of PLCs, specifically by displaying the presence or absence of continuity (power flow) within the rungs of ladder logic associated with the final control element of interest.

Configure a discrete device driver for each final control element. Initially, this appeared for distributed control systems (DCSs), but with the availability of subroutine facilities, is now possible with PLCs. Instead of programming the logic to support the state feedbacks, a discrete device driver is configured for each final control element. To support the minimal list of features presented earlier, the following parameters must be specified:

1. Output configuration for discrete outputs to the final control element
2. Hardware address of the discrete output to the final control element
3. Sense mode
4. Input address of state 0 feedback, if present
5. Input address of state 1 feedback, if present
6. Transition time.

4.3.5. Valves on a Piping Header

Figure 4.9 illustrates a piping header that can deliver material to two vessels. The amount of material transferred is determined from a single flow measurement located on the common header. This arrangement imposes the restriction that material can be transferred to either vessel, but not both at the same time.

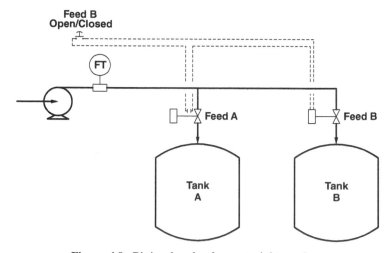

Figure 4.9. Piping header for material transfers.

Such arrangements are common in batch facilities. To make sure that all material passing through the flow meter is being transferred to the desired destination, the following must be assured:

1. While material is being transferred to **Tank A**, valve **Feed B** (the feed valve to **Tank B**) must be closed.
2. While material is being transferred to **Tank B**, valve **Feed A** (the feed valve to **Tank A**) must be closed.

Figure 4.9 illustrates conceptually what is required. A latching or maintained push button is installed so the operator can command valve **Feed B** to open. The diagram in Figure 4.9 illustrates two requirements:

1. The command to open valve **Feed B** must somehow "pass through" the limit switch for state **Closed** on valve **Feed A**. This can only happen if the limit switch is closed.
2. If the limit switch on valve **Feed A** is not closed, valve **Feed B** must be closed.

It is not sufficient to assure that valve **Feed A** is closed at the start of the material transfer to **Tank B**; valve **Feed A** must remain closed throughout the transfer. Should valve **Feed A** open during a transfer, valve **Feed B** must immediately close.

Prior to the advent of digital controls, the field wiring was arranged so that these requirements were met. Except in applications where personnel safety is involved, this practice has largely disappeared. Without getting into the details of wiring practices, the effective result was similar to that illustrated in Figure 4.9. Valve **Feed B** opens on continuity in the output circuit. The limit switch on valve **Feed A** is wired so as to be in series with the output circuitry for opening valve **Feed B**. State 0 (no continuity) on the hardware output means that valve **Feed B** is closed, so wiring the limit switch on valve **Feed A** in series means that valve **Feed B** will be closed whenever the limit switch on valve **Feed A** is open.

With a discrete device driver, the approach illustrated in Figure 4.10 is often supported. The discrete driver accepts a discrete input that serves as a permissive whose function is as follows:

True. Final control element is permitted to be in the state that corresponds to a hardware output of 1.

False. Final control element must be in the state that corresponds to a hardware output of 0. Normally this is the safe state.

Naturally, implementations that behave in the opposite manner are also possible; that is, a logic input of **True** forces the field device to its safe state.

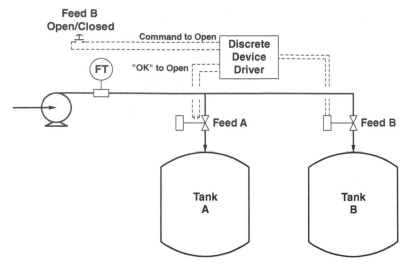

Figure 4.10. Discrete device driver with permissive to allow state open.

Yet another possibility is a feature often referred to as **Forced Tracking**. Two inputs are required to the discrete device driver:

1. Input to activate forced tracking. When forced tracking is active, the field device is driven to the state specified by the second input.
2. Input to specify desired state of the field device when forced tracking is active. If forced tracking is not active, this input has no effect.

For the discrete device driver for valve **Feed B** illustrated in Figure 4.10, the input from the limit switch for state **Closed** for valve **Feed A** can be used for the permissive. Most applications are more complex. Instead of using an input from the field, logic within the controls computes the true–false value of the permissive.

4.3.6. Ignoring a Limit Switch

Although installing the limit switches for state feedbacks within the actuator housing has greatly enhanced the reliability, failures still occasionally occur. If the plant manager arrives in the morning to discover that production has been stopped for 6 hours because of a problem with a limit switch, he or she will not be amused.

Preferably, the process controls provide mechanisms for ignoring a limit switch. Otherwise, production personnel find ways to defeat them. In the era of conventional wiring, attaching jumpers (often called "alligator clips") as illustrated in Figure 4.11 effectively bypassed the limit switch on the **Closed** state for valve **Feed B**. Such procedures were rarely, if ever, officially

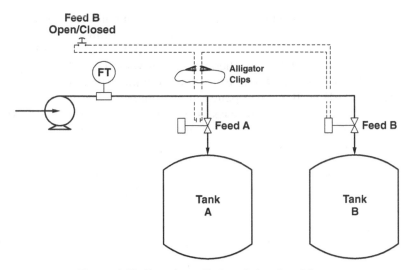

Figure 4.11. Ignoring a limit switch—the old way.

sanctioned, but were effectively tolerated for the sake of "getting the pounds (or kilos) out." Unfortunately, alligator clips were added but not always removed. Such practices have the potential to degrade the discrete logic accompanying outputs to field devices to the point that the logic ceases to be effective.

With discrete device drivers, jumpers such as alligator clips cannot be added without some side effects. For the limit switch on the **Closed** state for valve **Feed A**, adding an alligator clip will permit valve **Feed B** to open. However, when valve **Feed A** is opened, the presence of the alligator clip will result in an **Invalid** state for valve **Feed A**. At a minimum, the **Invalid** state causes an alert to be issued to the process operator. But being considered a failure, any control actions that involve valve **Feed A** are suspended, and any affected final control elements are driven to their safe states.

Most discrete device drivers permit authorized personnel to instruct the discrete device driver to ignore either or both state feedbacks. This approach has several advantages:

1. The need to add jumpers such as alligator clips to the field wiring is eliminated.
2. The commands to ignore a limit switch can be restricted to more knowledgeable personnel such as shift supervisors.
3. The time and date that the command to ignore a limit switch was executed can be captured.
4. The system can generate a report listing all state feedbacks that are currently ignored.

In the same sense that alligator clips are added but never removed, limit switches can remain ignored for an extended period of time. This defeats the purpose for which the state feedback was installed. For the piping header in Figure 4.9, the limit switch is installed to prevent valve **Feed B** from opening when valve **Feed A** is not **Closed**. But while the limit switch on valve **Feed A** is ignored, valve **Feed B** can be opened at any time.

The need to ignore limit switches is essential, but must be used wisely. Encouraging its wise use has led to several variations in the ignore command, including the following:

1. The ignore command enables only a single operation of the final control element. That is, after each command is executed, the ignore is automatically cleared.
2. For the final control elements used in the manufacture of a batch, all ignores are cleared at the start of the batch (or at some other appropriate point during the production of the batch).
3. The ignores remain in effect until removed by authorized personnel.

Basically, the objective is to get the priority of repairing a limit switch sufficiently high that it actually gets done.

With the capability to ignore a limit switch incorporated into the discrete device driver, a slight modification to the configuration of the permissive in Figure 4.10 should be considered. The discrete device driver for valve **Feed A** maintains a value for the state of valve **Feed A** as **Closed**, **Open**, **Transition**, or **Invalid**. When configuring the discrete device driver for valve **Feed B**, there are two options for the permissive:

1. The limit switch on state **Closed** for valve **Feed A** must be closed.
2. The state of valve **Feed A** as maintained by the discrete device driver must be **Closed**.

When an ignore is applied to a bad limit switch on state **Closed** for valve **Feed A**, the value of the state from the discrete device driver will be **Closed**, which permits the material transfer to **Tank B** to proceed.

4.4. ASSOCIATED FUNCTIONS

Under certain conditions, some final control elements must be in a specified state, regardless of the state prescribed by the process controls. Three examples are the following:

1. Interlocks
2. Emergency stop
3. Local or maintenance mode.

Each will be discussed in detail shortly.

For a given final control element, each of these can be implemented in two ways:

1. By incorporating the required function within the process controls, which may necessitate additional inputs and/or logic. The output from the process controls always determines the state of the final control element.
2. By incorporating the function into the field wiring, possibly with additional hardware. In effect, the function is inserted between the output(s) of the process controls and the final control element. Under certain conditions, the output from the controls does not determine the state of the final control element.

The issues pertaining to the selection of the approach to use will be discussed shortly.

When panels were routinely installed, another function had to be included in the previously mentioned list—a remote/local switch. In continuous processes, a remote/local switch was usually installed for each individual final control element. But in a batch facility, a single remote/local switch usually pertained to several final control elements, either all final control elements for a panel or all final control elements for a unit of equipment. The desire was for the unit to either be controlled by the process operator or controlled by the process controls. Controlling some by the operator and others by the controls was generally viewed as not a good idea.

Remote/local switches require tracking and similar functions much like those required for the other functions in the list. However, elimination of a panel also eliminates the need for one or more remote/local switches. Panels are occasionally installed for certain items of equipment, so remote/local switches have not entirely disappeared. But their use has dropped to the point that they will not be included in the discussion herein.

4.4.1. Tracking

When interlocks, emergency stops, and so on, are implemented external to the process controls, tracking can provide the following:

1. The state of the final control element can be accurately displayed to the process operator.
2. When functions such as an emergency stop are determining the state of a final control element, attempts by the process operator to change the state of the final control element can be refused. The state is being determined by the external logic, so actions by the process operator have no effect.
3. Any logic within the process controls that changes the state of the final control element can be suspended or terminated.

4. Some logic requires that the final control element be in a given state. An accurate value for the state is required to assure that this logic can continue.

5. The transition from the state being imposed by the external logic to the state being driven by the process controls must be executed smoothly. Motors should not abruptly start or stop, valves should not suddenly open or close, and so on.

Providing tracking requires at least one additional discrete input and sometimes two:

1. A discrete input is required to indicate to the process controls when an interlock is being imposed, when an emergency stop is in effect, when the final control element is being operated locally, and so on. At these times, tracking should be active.

2. An indication of the current state of the field device is required. Usually this can be determined from the discrete inputs provided for the state feedbacks. Sometimes it can be assumed that the final control element is in its safe state, which usually corresponds to a value of 0 for the output to the final control element. But occasionally, an additional discrete input must be added to the process controls that indicates the state of the final control element.

When tracking is active, two actions are performed:

1. The current state of the final control element is either determined from the inputs provided for this purpose or assumed to be in its safe state. When two state feedbacks are provided, it is possible that conflicting information is provided. The transition time must applied to allow for the device changing state. Thereafter, the state of the final control element is indicated as **Invalid**.

2. The states of the hardware discrete outputs must be set consistent with the state of the final control element. If the state is **Invalid**, the states of the hardware discrete outputs are set to the safe state. If the state is **Transition**, the states of the hardware discrete outputs are not changed.

The tracking capability is usually incorporated into the discrete device driver. When tracking is active (as based on the input for this purpose), the discrete device driver uses the state feedbacks to update the indicated state of the final control element, and then sets the hardware discrete output accordingly. If configured correctly, the discrete device driver does essentially all of the work required for tracking.

Configuring tracking into the discrete device driver avoids another problem that usually becomes more than a nuisance. Suppose the process controls have opened a valve. The hardware discrete output is set accordingly, and the state feedback(s) confirm that the valve is open. Then an emergency stop is acti-vated that forces the valve to close. Without tracking, the result is an **Invalid** state for the valve—the state feedbacks are not consistent with the hardware output(s) to the valve. The **Invalid** state is normally understood to be a failure in the valve itself, but in this case, is the result of the emergency stop. Not only is the alert generated as a result of the **Invalid** state misleading, it falls into the nuisance alarms category. These are distractions to operations personnel, and efforts must be directed to eliminating (or at least greatly reducing) such alarms.

4.4.2. Interlocks

For some final control elements, a specific state must be imposed should certain conditions arise within the process. How these are implemented usually depends on the potential consequences, the extremes being the following:

Minor. Examples include equipment damage that can be quickly repaired, just making a mess by overflowing a vessel, or similar. These are "process interlocks"; the interlock logic is usually incorporated into the process controls.

Major. This is definitely the case if personal injury is a potential conse-quence. These are "safety interlocks," so most insist on external imple-mentations of the interlock logic, either as hardwired implementations or PLC-based implementations.

Many situations fall between these two extremes. Where one draws the line between implementing within the process controls and implementing exter-nally varies from one organization to another.

When the interlock logic is implemented within the process controls, it is known when the interlock is imposed. The process controls change the state of the output accordingly, and logic can be incorporated to provide a smooth transition back to normal operations.

But when the interlock logic is implemented externally, how do the process controls know that the interlock is being imposed? The external interlock logic must provide a discrete input to the process controls that indicate that the interlock is being imposed. This serves several purposes:

1. Inform the process operator that the interlock is being imposed.
2. The process operator should be prevented from changing the state of any affected final control elements. Their states are being imposed by the external interlock logic, so operator actions on these final control ele-ments have no effect.
3. Tracking can be initiated on the affected final control elements so that the return to normal control is smooth.

4.4.3. Emergency Stop

These are most commonly installed for rotating equipment, the objective being to quickly stop the equipment should a hazardous situation of any type arise. An emergency stop can be specific to one field device (usually an item of rotating machinery), or an emergency stop can affect more than one field device. As an example of the latter, a single emergency stop associated with a material transfer may close valves, stop pumps, and so on.

The emergency stop is usually a button that is physically located within the proximity of the equipment to be stopped. Continuity in the circuit containing the emergency stop button allows the equipment to continue running. Pressing the emergency stop button breaks the continuity in the circuit. The emergency stop must be latched, either mechanically within the emergency stop push button (it remains depressed until released) or by associated latching circuitry (a separate push button is required to reset or clear the emergency stop).

Emergency stops for material transfers are often implemented within the process controls, but only when no personnel safety issues arise. Only when the consequences are minor would an emergency stop be incorporated into the logic within the process controls.

When the emergency stop pertains to rotating equipment, personnel safety is often an issue. If so, the emergency stop will be implemented via hardwired field wiring or possibly in a PLC. Figure 4.12 illustrates an emergency stop that is incorporated into the motor starter circuitry. The push button is equipped with a mechanical latching mechanism. Once depressed, it remains depressed until the latching mechanism is released.

Figure 4.12 is deficient in one regard. Depressing the emergency stop has the desired effect—the agitator motor stops. The state feedback to the

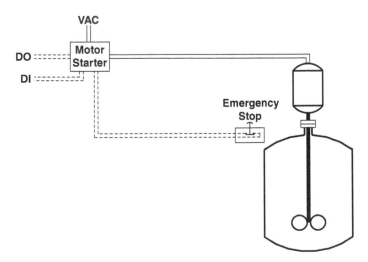

Figure 4.12. Emergency stop for a motor.

process controls indicates that the motor has stopped, which causes the discrete device driver to issue an alert indicating that the state of the field device is not the expected state. However, this alert is misleading. Such alerts from the discrete device driver are usually understood to be the result of some problem within the field device. On displays to the process operator, the state of the agitator motor is indicated as **Invalid**—the state feedback does not agree with the state corresponding to the outputs from the process controls.

But in this case, there is no problem with the final control element. Instead, the motor stopped because the emergency stop was depressed. To provide accurate information to the process operator, a discrete input must be installed to indicate that the emergency stop has been depressed. This input is used as follows:

1. Generate an alert indicating that an emergency stop is in effect.
2. Initiate tracking by the discrete device driver. The displayed agitator state is accurate. The **Invalid** state is not indicated, and no alert is issued to the process operators.

This input must change state when the emergency stop is imposed and must remain in this state until the emergency stop is reset or cleared. This discrete input serves the same purposes as described earlier for an interlock.

4.4.4. Local or Maintenance Mode

These are justified for reasons such as the following:

1. Operations such as cleaning are often not automated. Providing a capability in the field to manually operate valves, motors, and so on, can shorten the time required to perform such operations.
2. Performing maintenance on rotating equipment may require the capability to start a motor, stop a motor, or "jog" a motor.

The alternative to a local capability is for the field personnel to contact the control room personnel and ask them to operate the field devices. For infrequent activities, this is usually acceptable, but not for routine activities that are frequently performed.

Figure 4.13 illustrates a local panel that permits maintenance personnel to start and stop an agitator motor. This panel provides a three-position switch that provides the following options:

1. Start the motor. Mode is local.
2. Stop the motor. Mode is local.
3. Operate the motor via the process controls. Mode is remote.

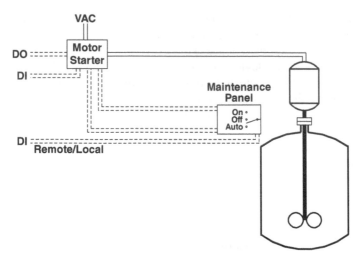

Figure 4.13. Maintenance panel for operating a motor.

An essential component of the interface is a discrete input to the controls that indicate the mode, that is, local or remote. An additional input may be required depending on how the local actions (motor start and stop) are implemented:

1. Within the motor starter circuitry that is external to the process controls. The local/remote discrete input indicates the mode of operation; the state feedback from the motor starter circuitry provides the running/stopped indication. The configuration in Figure 4.13 reflects this implementation.
2. Within the process controls. An additional discrete input is required to indicate the desired state of the motor (running or stopped). During the local mode of operation, the process controls respond by setting the output to provide the desired state of the motor. During the remote mode of operation, this input is ignored.

When the motor is in the local mode, the process operator must certainly be informed that the motor is on local and any attempts by the process operator to change the state of the motor must be refused. If the motor start/stop function is incorporated into the external motor starter circuitry, tracking is usually desirable to perform the same functions as described previously for interlocks. When maintenance is being performed, an accurate indication of the device state may not be essential, but all alerts associated with the device must be suppressed.

4.5. BEYOND TWO-STATE FINAL CONTROL ELEMENTS

Most of the final control elements encountered in process facilities are two-state (discrete) devices, possibly equipped with one or more state feedbacks.

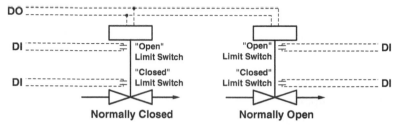

Figure 4.14. Dual valves.

But occasionally, more complex arrangements are encountered. This section discusses two possibilities:

1. Two (or more) two-state devices are operated in a specified fashion, usually by the same hardware output.
2. A few final control elements have more than two states.

4.5.1. Dual Valves

Figure 4.14 illustrates an arrangement that is occasionally encountered in process facilities. Two valves are installed, one of which must be open at all times. To assure this, the following approach is used:

1. One valve (**Valve A**) is normally closed (fail-closed); the other valve (**Valve B**) is normally open (fail-open).
2. A single output from the controls drives both valves.

The state of the hardware output has the following effect:

Hardware Output State	Valve A	Valve B
0	Closed	Open
1	Open	Closed

In Figure 4.14, both valves are equipped with limit switches to provide state feedbacks. Configuring a two-state valve point for each valve allows the discrete device driver to resolve the state feedbacks individually for each valve.

Suppose the hardware output(s) from the valve point for **Valve A** drives both field valves. Nothing is connected to the hardware output from the valve point for **Valve B**. But for the discrete device driver for **Valve B** to correctly process the state feedbacks for **Valve B**, the field device state for **Valve B** must always be the opposite of the field device state for **Valve A**. In essence, **Valve B** is slaved to **Valve A**, and the logic to do so must be provided within the controls.

4.5.2. Three-State Devices

Final control elements with more than two states are occasionally encountered. Here are two examples:

1. Two-speed agitator motor. The states are as follows:
   State 0—**Stopped**
   State 1—**Run Slow**
   State 2—**Run Fast**.
   A reversible motor also has three states:
   State 0—**Stopped**
   State 1—**Run Forward**
   State 2—**Run Reverse**.
2. Dribble flow. This is used to improve the accuracy of a material transfer. The bulk of the transfer is at a high flow rate, but near the end of the transfer, the flow rate is greatly reduced so that a more precise cutoff can be achieved. Some implementations utilize two two-state valves, a large one for the main flow and a small one for the dribble flow. There are four possible states, but two are essentially the same. When the large valve is open, the small valve contributes very little to the total flow, so whether it is open or closed is inconsequential. Herein the following states will be defined:
   State 0—**Closed** (both valves closed)
   State 1—**Dribble Flow** (dribble valve open; main valve closed)
   State 2—**Full Flow** (both valves open).

A higher number of states are possible. For example, a two-speed reversible motor has five states: **Stopped**, **Run Slow Forward**, **Run Fast Forward**, **Run Slow Reverse**, and **Run Fast Reverse**.

State feedbacks are usually provided, but not necessarily for all states. However, it is possible to have more state feedbacks than states. Consider the dribble flow arrangement described earlier. If both valves (full flow valve and dribble flow valve) are both equipped with two limit switches, there are four state feedbacks but only three states. For this case, it is probably better to configure a two-state valve point for each valve.

5

MATERIAL TRANSFERS

Batch operations involve numerous material transfers, all of which must be executed expeditiously and accurately. The typical consequences of not doing so are the following:

1. Any delay in completing a material transfer often lengthens the time to manufacture a batch of product.
2. Transferring less than or more than the specified quantity for the material is likely to adversely impact product quality.

Material transfers of various types are common within batch facilities:

1. From storage to a process vessel
2. From one process vessel to another
3. From a process vessel to storage.

A material transfer may be for a liquid, a solid, a gas, or a two-phase media (slurry or emulsion).

Material transfer systems generally fall into the following categories:

1. Single-source, single-destination; basically equipment is dedicated to the material transfer
2. Single-source, multiple-destination

Control of Batch Processes, First Edition. Cecil L. Smith.

3. Multiple-source, single-destination
4. Multiple-source, multiple-destination.

Most batch facilities are designed with some combination of these.
    The choice of the approach to use involves two considerations:

Equipment cost. Installing dedicated equipment for each material transfer
    is prohibitively expensive. The lowest cost is usually a material transfer
    system capable of transferring a material from one of several sources to
    one of several destinations.

Production delays. A material transfer system for multiple sources and/or
    multiple destinations can execute only one material transfer at a time.
    Although judicious batch scheduling can minimize the impact, delays in
    production operations arise whenever one material transfer cannot com-
    mence until another material transfer has completed.

The optimum design of the material transfer system(s) for a batch facility
involves balancing the cost of the material transfer equipment and the productivity of the production equipment. Discrete simulation is very effective
for establishing the plant productivity given the following:

1. Configuration of the material transfer equipment
2. Slate of products to be manufactured in the facility.

For specialty batch facilities, the latter is a major issue. The number of products
is large and is continually changing. The products being manufactured a year
from now are unlikely to be the same as today. The material transfer equipment must not only be efficient, but it must also be flexible.
    The design basis for most batch facilities is that simultaneous material
transfers will not occur. But as will be discussed in the last section of this
chapter, this may be reconsidered once the plant is in operation.

5.1. MULTIPLE-SOURCE, SINGLE-DESTINATION MATERIAL TRANSFER SYSTEM

Figure 5.1 illustrates a material transfer system capable of transferring a liquid
raw material from one of several sources to a single destination. Such systems
are commonly installed for charging raw materials to production vessels, with
separate systems provided for each production vessel.

5.1.1. Key Characteristics

The key characteristics of the material transfer system in Figure 5.1 are as
follows:

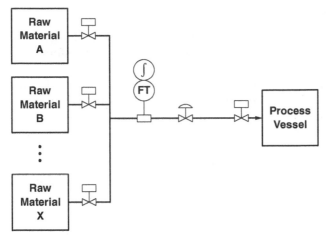

Figure 5.1. Multiple-source, single-destination raw material transfer system.

1. A block valve must be installed for each source vessel. Logic must be provided to assure that only one of these valves can be open at any instant of time.
2. A single block valve must be installed for the destination vessel.
3. The driving force for fluid flow may be any of the following:

   Gravity flow. The raw material tanks must be elevated relative to the destination vessel.

   Vacuum. Evacuating the destination vessel can provide up to 1 atm pressure differential for fluid flow.

   Pressure. Pressurizing a raw material storage vessel is possible only if the storage vessel is rated as pressure vessel.

   Pump. Usually a single pump is installed upstream of the flow meter.
4. If the flow rate must be controlled during any material transfer, a control valve must be installed.

5.1.2. Metering Issues

If the amount of material transferred is to be obtained by totalizing the flow, a coriolis meter is normally the preference for the flow meter illustrated in Figure 5.1. In addition to being a mass meter, a couple of other issues favor the coriolis meter:

1. A variety of materials pass through the meter. The coriolis meter is not affected by fluid properties such as viscosity, density, and so on.
2. The flow rates can differ widely, making the wide turndown ratio of the coriolis meter an advantage.

5.1.3. Purging

After most transfers, the piping in Figure 5.1 must be purged. The only exception is when mixing the materials for two successive transfers causes no adverse consequences.

Although a liquid or an inert gas is sometimes required, compressed air is most commonly used to "blow the line dry." Although not a problem for a coriolis meter, this practice cannot be permitted for a turbine meter.

Purging transfers the contents of the piping to one of the following:

Destination vessel. The volume of the piping in the header is known. The amount of material in the header can be computed and added to the amount of material transferred to the destination vessel. This imposes a lower limit on the amount of the transfer.

Source vessel. This raises the possibility of contamination in the source vessel.

Waste or holding tank. The issue is the final destination of this material. Sometimes it can be used elsewhere in the facility, but if not, costs are incurred to dispose of this material in addition to the purchase price of the material lost.

5.1.4. Impact on Production Operations

The following statements apply to the material transfer system in Figure 5.1:

1. For the raw materials that the material transfer system can deliver, only one raw material transfer can be in progress at a point in time.
2. While the transfer of a given raw material is in progress to one production vessel, the same raw material can be transferred to another production vessel. This assumes the following:
   (a) A raw material transfer system of the type in Figure 5.1 is provided for each production vessel.
   (b) The change in the weight of the source vessel does not provide the basis for determining the amount of material transferred.

5.1.5. Solids

Figure 5.2 illustrates a multiple-source, single-destination raw material transfer system for solids. Eight bins form an octagonal arrangement. For each bin, a screw feeder can transfer the solids onto a weighing pan that is normally suspended via a single load cell. For each bin, the transfer is as follows:

1. The specified amount of the solid material is transferred from a bin to the weighing pan.
2. The weighing pan is emptied into the destination vessel.

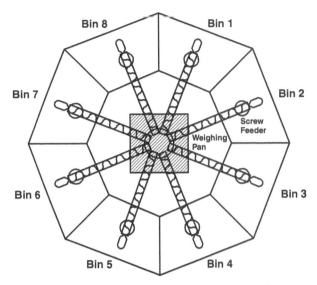

Figure 5.2. Multiple-source, single-destination material transfer system for solids.

Although some weighing systems execute the transfers in the same order as the physical arrangement of the bins, this is a software limitation, not a limitation imposed by the equipment in the field.

5.2. SINGLE-SOURCE, MULTIPLE-DESTINATION MATERIAL TRANSFER SYSTEM

Figure 5.3 illustrates a material transfer system capable of transferring a liquid material from a single source to any one of multiple destinations. Such configurations are not only commonly used for products, but can also be used for raw materials.

5.2.1. Key Characteristics

The important characteristics of material transfer system in Figure 5.3 are as follows:

1. A block valve must be installed for each destination vessel. Logic must be provided to assure that only one of these valves can be open at any instant of time.
2. A single block valve must be installed for the source vessel.
3. The driving force for fluid flow is usually a pump installed in the vicinity of the source vessel, but gravity flow, evacuated destination vessel, or pressurized source vessel are occasionally encountered.

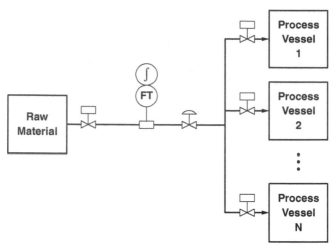

Figure 5.3. Single-source, multiple-destination raw material transfer system.

    4. If the flow rate must be controlled during any material transfer, a control valve must be installed.

5.2.2. Metering Issues

A coriolis meter is normally chosen for the flow meter illustrated in Figure 5.3, but mainly because it is a mass meter. As compared with a material transfer system for multiple sources (as in Figure 5.1), the advantages are less significant:

    1. A single material passes through the meter. Temperature and possibly composition affect physical properties of the fluid, but differences will be much less than when different materials pass through the meter.

    2. The flow rates may differ from one batch to the next, but the differences are usually much less than when different materials pass through the same meter. In most cases, a smaller turndown ratio is adequate, which makes the wide turndown ratio of the coriolis meter less of an advantage.

5.2.3. Purging

Although purging is usually not required, the following issues occasionally arise:

    1. Between transfers, the pump is stopped. With time the piping usually "bleeds back" and can admit air into the flow meter. Most coriolis meter manufacturers provide a "low-density cutoff" which forces the flow meter

output to zero if the density of the material within the flow meter is less than a specified value.

2. Some materials cannot be allowed to stagnate in the piping. Some change characteristics, through reactions such as polymerization. Others will "set up" if flow is stopped for even a short period of time. For example, a starch slurry can readily be pumped provided a continuous flow is maintained, but should the flow stop, the starch begins to set up very quickly. All transfers must be immediately followed with a water flush without permitting the flow to stop at any time.

5.2.4. Impact on Production Operations

When used for a raw material, only one material transfer system of the type in Figure 5.3 is typically installed in the plant for that raw material. Consequently, only one transfer can be in progress for that raw material at a given instant of time. This definitely provides the potential to disrupt production operations. While a transfer of a given raw material is in progress to a production vessel, any other production vessel needing that raw material must wait until the current transfer is completed. Long-duration raw material transfers must be scheduled very carefully.

5.3. MULTIPLE-SOURCE, MULTIPLE-DESTINATION MATERIAL TRANSFER SYSTEM

Figure 5.4 illustrates a material transfer system capable of transferring a liquid material from one of several sources to any of several destinations. The main attraction of the multiple-source, multiple-destination configuration in Figure

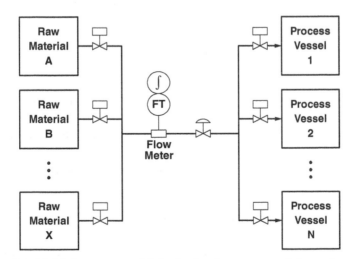

Figure 5.4. Multiple-source, multiple-destination raw material transfer system.

5.4 is its lower cost as compared with the approaches previously presented. But as will be discussed, the potential to disrupt production operations is high. Once the system begins to limit production capacity, higher-cost alternatives can usually be justified.

5.3.1. Key Characteristics

The key characteristics of the material transfer system in Figure 5.4 are as follows:

1. A block valve must be installed for each source vessel. Logic must be provided to assure that only one of these valves can be open at any instant of time.
2. A block valve must be installed for each destination vessel. Logic must be provided to assure that only one of these valves can be open at any instant of time.
3. The driving force for fluid flow is usually a pump.
4. If the flow rate must be controlled during any material transfer, a control valve must be installed.

5.3.2. Pneumatic Conveyers

Pneumatic conveying systems are often of the multiple-source, multiple-destination variety with structures similar to that in Figure 5.4. An air blower replaces the pump; diverter valves and rotary valves replace the block valves. Directly measuring the flow of the solids is impractical. The amount transferred must be based on the change in weight of the source or destination vessel, but in many cases, the transfer is continued until the source vessel is empty.

5.3.3. Hose Stand

A conventional implementation of the material transfer system in Figure 5.4 was the "hose stand." This system consisted of three components:

1. A transfer pump (and possibly a flow meter) with connections for flexible hoses on both suction and discharge.
2. Dedicated piping from each source vessel. Each pipe terminated with a connection for a flexible hose.
3. Dedicated piping to each destination vessel. A connection for a flexible hose was installed on the inlet to each pipe.

To transfer the desired raw material to a given production vessel, the operator did the following:

1. Connected the flexible hose on the pump suction to the pipe from the storage vessel for that raw material
2. Connected the flexible hose on the pump discharge to the pipe to the destination vessel.

This was an inexpensive way to provide the same capability as afforded by the two piping headers in Figure 5.4. But as automation progressed within batch facilities, hose stands have largely disappeared.

5.3.4. Metering Issues

A coriolis meter is normally chosen for the flow meter illustrated in Figure 5.3. The flow meter must be capable of metering each of the raw materials, often with widely different flow rates. The considerations are largely the same as presented previously for the multiple-source, single-destination configuration in Figure 5.1.

5.3.5. Purging

The purging issues are largely the same as for the multiple-source, single-destination configuration in Figure 5.1. The most common approach is to use compressed air to "blow the line dry." However, the multiple-source, multiple-destination systems are more challenging to adequately purge. The piping must be carefully arranged so that all liquid is removed, which may require admitting the purge fluid at more than one location.

5.3.6. Impact on Production Operations

For the multiple-source, multiple-destination transfer system in Figure 5.4, the potential to disrupt production operations is very high. If the source vessels contain raw materials and the destination vessels are production vessels, the limitations of the material transfer system in Figure 5.4 are as follows:

1. Only one transfer of a raw material from any of the source vessels can be in progress at a given time. All other process vessels needing any raw material must wait until the current transfer is completed.
2. If a transfer of any raw material to any production vessel is in progress, no raw material can be transferred to any other production vessel until the current transfer is completed.

Such a raw material transfer system can impose serious limitations on production operations.

5.4. VALIDATING A MATERIAL TRANSFER

The design basis is usually the following:

Primary measurement for amount transferred. The amount transferred must be based on the most accurate measurement available. The options are usually the following:
1. When using a flow meter, a coriolis meter is installed.
2. Change in weight of the source or destination vessel is often a possibility. But when a coriolis meter is also installed, this raises issues as to which is the most accurate. This was discussed in a previous chapter.

Secondary measurement for amount transferred. A second measurement of the amount transferred is used to detect gross errors in the primary measurement. The most common approaches to obtaining a second value for the amount transferred are the following:
1. Change in destination vessel weight (or liquid volume)
2. Change in source vessel weight (or liquid volume)
3. Additional flow measurement.

5.4.1. Flow Measurement Issues

The coriolis meter is usually the meter of choice for liquid material transfers. Use of other types of meters for the primary measurement is unusual, but other types are routinely used for the secondary measurement. Issues such as the following arise with the alternatives to the coriolis meter:

Magnetic flow meter. With no obstruction to fluid flow, the pressure drop is the same a straight pipe of the same diameter. The common practice of installing a meter with a tube diameter smaller than the line size may not be a good idea in gravity flow applications. However, the main obstacle is that the electrical conductivity of most organic fluids is below the minimum required by a magnetic flow meter.

Vortex shedding meter. This meter only works for turbulent flow—the sensed flow is zero if the flow is laminar. Especially for gravity flow, turbulent flow is not assured.

Ultrasonic flow meter. The small line sizes encountered in most batch applications become an issue. To obtain the necessary path length for the ultrasonic beam, multiple transits are required for small pipes. These are easily disrupted by buildups on the wetted surfaces inside the tube.

Turbine meter. Being accurate volumetric meters, these were commonly installed in batch facilities prior to the advent of the coriolis meter. Even though a variety of bearing assemblies appeared over the years, most operational problems with turbine meters pertained in some way to the bearings. And with these meters, procedures such as "blow the line dry" must be avoided.

All of the previously mentioned are volumetric meters. The options for converting between mass and volume are as follows:

Constant density. The transfer amount can be converted to volume. Alternatively, the measurement range of the volumetric meter can be expressed as mass.

Variable density. The volumetric flow must be converted to mass flow and then totalized. This approach permits temperature compensation to be provided for the liquid density.

The total amount transferred is obtained by totalizing the flow, which is easily implemented in digital controls.

5.4.2. Transfer Amount from Vessel Weight

If simultaneous transfers (either in or out) do not occur in a vessel, the change in weight (or volume) of the vessel can be used as the primary and/or secondary value for the amount transferred. The change in weight can be determined in several ways:

Weight of the vessel. This is generally the preferred approach.

Level in the vessel. This is the level from which the volume of liquid within the vessel can be computed. To convert volume to weight, a value for the density is required. Stratification is likely when the vessel contents are not agitated, which complicates obtaining a value for the liquid density. Temperature compensation for liquid density is easy to provide, but stratification also complicates obtaining a good value for the temperature.

Manual readings. This could simply be reading the level via a sight glass. It could be reading the liquid mark from a stick dipped into the vessel, which at one time was very common for buried storage tanks. Needless to say, such practices are disappearing. One possibility is to use the manual reading as a third check value. If the value of the primary measurement does not agree with the secondary measurement, operations personnel can obtain the manual reading to hopefully resolve which value is correct. Unfortunately, the manual value is usually the least accurate of all.

Installing a weight measurement on a process vessel is very common. When installing weight measurements on raw material and product storage tanks, obstacles such as the following can arise:

1. Structural considerations favor resting the bottom of large storage tanks on the ground.
2. Buried tanks (such as at gas stations) are occasionally encountered in industrial batch facilities.

5.4.3. Validation Logic

The objective of the following approach is to detect significant errors in the value for the total amount transferred:

1. Obtain a secondary measure for the amount transferred, that is, a value that does not rely in any way on the value sensed by the primary measurement.
2. At the end of the transfer, compare this value with the value from the primary measurement.

Should this value exceed some threshold, suspend further processing for this batch and alert operations personnel.

When both the primary and the secondary measurements are based on flows, the logic could be formulated in two ways:

1. Continuously compare the flows indicated by the two meters. Normally, logic must be incorporated to avoid shutdowns resulting from a short-term discrepancy in the two flows.
2. Totalize both flows and compare only at the end of the transfer, or possibly on more frequent intervals as will be described shortly.

The most common approach is to implement some version of the latter.

For long transfers, waiting until the end of the transfer to detect the discrepancy is too "after-the-fact." As the transfer proceeds, comparisons can be done on bases such as the following:

Time. On a specified interval (such as 30 minutes), compare the two values for the amount transferred.

Amount. On transferring a specified amount (such as 500 kg) based on the primary measurement for the amount transferred, compare with the value from the secondary measurement for the amount transferred.

The more frequent the two values for the change in weight are compared, the larger the contribution of the noise associated with weight measurements. The extreme would be to convert the change in weight to a flow. But as discussed in the chapter on measurements, converting change in weight to flow is a challenging endeavor.

5.4.4. Check Meter

Redundant flow measurements in batch plants consist of a primary flow meter and a secondary or "check" flow meter. However, this is not the customary redundant measurement application where either measurement can be used. No averaging, failover to the secondary meter, and so on, is performed. The only role of the check meter is to detect situations where the primary meter is not providing accurate values. All batch logic is based on the values derived from the primary meter.

Even for coriolis meters, problems can arise that cause significant errors in the measured flow. For example, coriolis meters are adversely affected by gas entrained in the liquid stream, which can be introduced by mechanical problems such as a leaking seal.

Such observations are the basis for the argument for using a different technology for the check meter. The desire is to detect anything that would cause the primary meter reading to be in error, not just a malfunction of one of the measuring devices. If two meters are based on the same principle, anything that affects one meter is likely to have the same effect on the other meter. By using a different technology in each meter, these effects are less likely to be the same, resulting in an indication that a problem exists.

When the primary meter is a coriolis meter, the desire that the check meter not rely on the same principle as the primary meter means that the check meter is almost certainly a volumetric meter. Although using meters such as magnetic, vortex shedding, ultrasonic, and the like are preferable, a head-type meter such as the orifice meter can be used. As the accuracy of the check meter decreases, a larger tolerance must be provided between the flow sensed by the primary meter and the flow sensed by the secondary meter.

When the secondary meter is a volumetric meter, two options are possible for comparing the values for the amount transferred:

1. Convert the volumetric flow from the secondary meter to a mass flow and compare mass flows.
2. Convert the mass flow from the primary meter to a volumetric flow and compare the volumetric flows.

The choice is usually dictated by the specification for the amount to transfer, which is usually in mass units.

5.5. DRIBBLE FLOW

The objective of all "dribble flow" configurations is to improve the accuracy of a material transfer. These can be applied to both liquid transfers and solid transfers.

5.5.1. Logic

The termination of a material transfer involves closing a valve, stopping a feeder, or something comparable. Such events do not occur instantaneously due to valve travel times and the like. The dribble flow concept has two objectives:

1. Execute the majority of the material transfer at as rapidly a rate as the equipment allows. Valves are fully open, feeders are running at maximum speed, and so on. This translates into the shortest possible time for the material transfer, which shortens the time required to manufacture a batch and results in a higher production rate from the plant.
2. Near the end of the transfer, transition to a slow flow rate called the "dribble" flow. Once termination of the material transfer is initiated by closing valves, stopping feeders, or the like, a much smaller amount of material will be transferred, thereby improving the accuracy of the material transfer.

This logic requires the following two values:

Preset. This is the target for the total amount of material to be transferred. The actual amount transferred should be as close to this value as possible.

Pre-preset. The transfer proceeds as rapidly as possible until this amount is transferred.

The material transfer rate transitions from full flow to dribble flow at the pre-preset.

The difference between the preset and the pre-preset must allow the transition to dribble flow to complete before the preset is attained. This value depends on the characteristics of the equipment, and should be a fixed amount (not a percentage of the amount to be transferred).

5.5.2. Two Block Valves

In the logic in Figure 5.5, the output from the flow totalizer is the input to two comparators:

Compare flow total to preset. The dribble flow valve is open provided the flow total is less than the preset.

Compare flow total to pre-preset. The main flow valve is open provided the flow total is less than the pre-preset.

In practice, the dribble flow is much less than the full flow. When the full flow valve is open, whether the dribble flow valve is open or closed is immaterial.

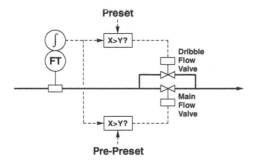

Figure 5.5. Dribble flow using two block valves.

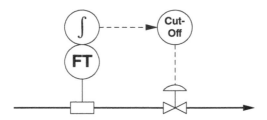

Figure 5.6. Dribble flow using a positioning valve.

In a few implementations, this logic is incorporated into the totalizer (or integrator) block. Such blocks provide three outputs:

TOTAL. Value of flow total.

Q1. Flow total is less than the preset. This output drives the dribble flow valve in Figure 5.5.

Q2. Flow total is less than the pre-preset. This output drives the full flow valve in Figure 5.5.

5.5.3. Single Positioning Valve

In the configuration presented in Figure 5.6, the output from the flow totalizer is the input to a two-stage cutoff block. The output of this block determines the opening of the positioning valve as follows:

Flow total less than pre-preset. The positioning valve is 100% open, giving the maximum flow rate.

Flow total between pre-preset and preset. The opening of the positioning valve is reduced substantially to give the dribble flow. The value of this opening is essentially a tuning parameter.

Flow total greater than preset. The positioning valve is fully closed.

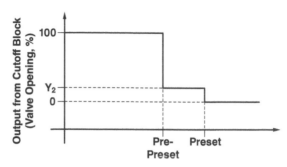

Figure 5.7. Relationship for cutoff block in dribble flow configuration.

The relationship for the cutoff block is illustrated in Figure 5.7 and is expressed as follows:

$$Y = \begin{cases} Y_1 & \text{if } X < \text{Pre-Preset} & \text{Full flow} \\ Y_2 & \text{if Pre-Preset} \leq X < \text{Preset} & \text{Dribble flow} \\ Y_3 & \text{if } X \geq \text{Preset} & \text{Flow stopped,} \end{cases}$$

where

Y = Output of cutoff block
Y_1 = Valve opening for full flow, normally 100%
Y_2 = Valve opening for dribble flow
Y_3 = Valve opening for no flow, normally 0%

The value of Y_2 is essentially a tuning parameter, and depends on the desired dribble flow rate and the valve characteristics. One advantage of this approach is that the value of Y_2 can be easily changed. In the configuration in Figure 5.5 that uses block valves, changing the value of the dribble flow requires changing the size of the dribble flow valve.

5.5.4. Dribble Flow for Solids

Solids handling is best left to those with considerable experience, hopefully with the materials being handled. The screw feeders illustrated in the system in Figure 5.2 work well for some materials, but not for all. Some materials tend to clump or bridge, and will not flow into the screw feeder. Other materials (powered limestone is one example) will easily fluidize, and when this occurs, material flows freely through the screw feeder even when it is stopped.

One impact is how dribble flow is implemented. Figure 5.8 presents three versions:

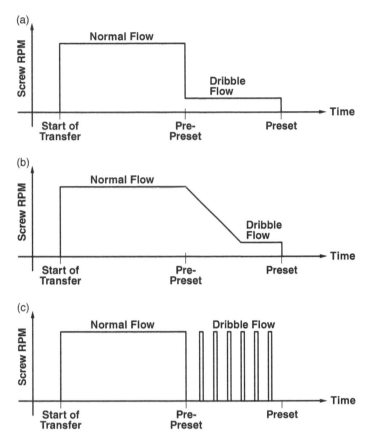

Figure 5.8. Dribble flow alternatives for solids: (a) abrupt transition to dribble flow; (b) gradual transition to dribble flow; and (c) dribble flow by jogging the feeder.

Abrupt transition from full flow to dribble flow. This is not advisable for materials that fluidize easily.

Gradual transition from full flow to dribble flow. The ramp rate is an adjustable parameter.

Jogging the screw feeder. Some materials will not move through the screw feeder when the speed is slow. For these, dribble flow is implemented by "jogging" the screw feeder motor. On each "jog," the screw rotates a specified number of degrees (also an adjustable parameter), which should discharge a relatively fixed amount of material into the weighing pan. The amount transferred in each "jog" determines the accuracy with which the material transfer can be executed.

Most systems of the type illustrated in Figure 5.2 permit a bin to be emptied, cleaned, and filled with a different material. The control software maintains a set of configurable parameters for each bin, one obvious entry being the name of the material. Since the dribble flow logic depends on the material, these configuration parameters must also provide options for the dribble flow. When the contents of a bin are changed, different values must be specified for these configurable parameters.

5.6. SIMULTANEOUS MATERIAL TRANSFERS

In the initial design of batch facilities, a common basis is that simultaneous material transfers will not be allowed. However, simultaneous material transfers can quickly become an issue. The day the plant manager has orders that cannot be filled, this restriction is very likely to be reconsidered. Most opportunities for simultaneous material transfers pertain to raw materials, but occasionally, opportunities arise for products.

To what extent simultaneous material transfers are possible depends largely on the material transfer systems installed in the plant. Consider the following two material transfer systems:

1. A single-source, multiple-destination material transfer system as in Figure 5.3 capable of transferring raw material A to a production vessel
2. A single-source, multiple-destination material transfer system as in Figure 5.3 capable of transferring raw material B to the same production vessel.

The raw material transfer systems are capable of simultaneously transferring raw material A and raw material B to the production vessel, but are subject to one restriction: The weight of the destination vessel cannot be used as the basis for determining the amount of either raw material that is transferred. If a secondary measurement of the amount transferred is required, there are two options:

1. Use the weight of the respective source vessels.
2. Install a second flow meter, preferably of a type different from the primary. Meters are required for both raw material transfer systems.

Actually, there is a third option—dispense with the secondary measurements for the amount of material transferred. The primary flow meter is likely to be a coriolis meter. These have an excellent track record for reliability. A plant manager with orders that cannot be filled has an incentive to trust a coriolis meter. The economic trade-off is relatively simple:

1. If the coriolis meter has a problem, the most likely consequence is that a product batch is off-spec. A cost can be established for this.

2. By executing simultaneous material transfers, the time to manufacture a batch can be reduced. From the reduction in this time, the increase in plant productivity can be calculated and translated into an economic benefit.

From these, one can compute a time interval required for the increase in plant productivity to offset the cost of a lost batch. If this interval is short, simultaneous material transfers are irresistible to a plant manager.

This analysis is purely based on economics. Occasionally, the consequences of an error in charging can be very serious, such as runaway reactions that can lead to a hazardous situation. In these cases, a secondary measure of the amount transferred must be retained. But if the economic incentive is sufficient, installation of secondary flow measurements can be justified.

If simultaneous material transfers were contemplated at the time the plant was constructed, a different arrangement of material transfer systems would likely be installed in the plant. To date, there seems to be little movement in this direction. In most plants, the major benefits of simultaneous material transfers can be realized with modest modifications to the production equipment. Often the major modification is to the control logic.

5.7. DRUMS

In specialty batch plants, the number of raw materials is large. Except for the major raw materials, dedicating a process vessel to a raw material is impractical. Many of the raw materials are stored in drums, which creates an incentive to charge some raw materials directly from drums. The approach is generally as follows:

1. A platform scale provides the weight measurement on which the raw material transfer is based.
2. The driving force for fluid flow is often provided by evacuating the process vessel. The drum is opened and a "dip tube" is inserted. However, pumping arrangements are possible.
3. A block valve is opened to start the transfer and closed to terminate the transfer.

The basic control logic is as follows:

1. Record the initial weight.
2. Subtract the amount to be transferred from the initial weight to obtain the target weight.
3. Open the transfer valve until the target weight is attained.

Unfortunately, there are a couple of complications:

1. For some transfers, the drum is emptied before the target weight is attained. The logic must provide for the transfer to span two drums.
2. For some transfers, the transfer amount exceeds the capacity of a single drum. The transfer will always involve two drums, but could span three (or possibly more).

A major concern in all specialty batch facilities is charging the incorrect material. Certainly the batch will be lost, but the potential exists for more severe consequences that are difficult to know in advance. A runaway reaction could overpressure the vessel, with no assurance that the relief devices are capable of coping. The products of the reaction could make cleaning the vessel very difficult, resulting in considerable lost production time.

Charging from drums definitely offers opportunities for errors with such consequences. How to assure that the material being charged is the desired material? Possibilities include the following:

Barcodes. Scanning may be for the material code and/or lot number (useful for tracking). Reading traditional barcodes in a production environment is unreliable. Alternatives such as magnetic barcodes and memory chips are better, but not entirely without problems.

Physical properties. For transfers from drums, basing the amount of material transferred on a coriolis meter instead of a platform scale offers an additional advantage. A coriolis meter can measure density in addition to flow. The density of the material being transferred can be compared with the expected density. Inexpensive measurements for electrical conductivity, refractive index, and so on, can be used in a similar fashion. Doing so increases the confidence that the correct material is being transferred, but falls short of absolute assurance. Composition analyzers could do the latter, but with significant additional cost, complexity, and production delays (sampling analyzers like chromatographs are not instantaneous).

Does the reduction in errors justify the additional expense and complexity incurred? Everyone seems to have an opinion, but in the end, they are just that—opinions. Generally, those who have experienced the consequences of charging the wrong material are in favor or doing whatever is practical.

6

STRUCTURED LOGIC FOR BATCH

Despite its definition in Wikipedia as a "highway in the southeastern part of the state of Sao Paulo in Brazil," in the process industries SP88 is generally understood to be a standard developed by the SP88 standards committee of the International Society for Automation (ISA; formerly known as the Instrument Society of America) that produced the document "Batch Control Part 1: Models and Terminology" [1]. The initial version appeared in 1995 and was updated in 2010. The SP88 committee has produced four additional documents pertaining to various aspects of batch control.

Founded at Leverkusen, Germany, in 1949 to represent the interests of the users of measurement and control technology in the chemical industry, Normenarbeitsgemeinschaft für Meß- und Regeltechnik (NAMUR) in der Chemischen Industrie changed its name in 1996 to Interessengemeinschaft Automatisierungstechnik der Prozessindustrie. In 1993, its working group on Recipe-Based Operation and Plant Control issued a document titled "Anforderungen an Systeme zur Rezeptfahrweise," which translates to "Requirements to be met by systems for recipe-based operations" [2].

In a sense, the main contribution of NAMUR's document was to get across to the control system suppliers the concept of a recipe as used in the specialty batch chemical industries. At the time, the control system suppliers understood the recipe to be a set of numbers, such as raw material amounts, temperatures, and so on. By using one or more of these numbers to designate options, this approach could be effectively applied to automating facilities such as pulp

Control of Batch Processes, First Edition. Cecil L. Smith.
© 2014 John Wiley & Sons, Inc. Published 2014 by John Wiley & Sons, Inc.

digesters. However, applying this approach in the specialty chemical industries proved to require too many options, resulting in extreme complexity. The message to the suppliers: a recipe in the specialty chemical industry includes the product-specific logic as well as a formula. For example, in addition to the amounts to be added, the recipe includes the order in which the materials must be added.

The process industries in general, and the ISA in particular, are enamored with standards. However, the effectiveness of standards, especially for anything related to computers, is questionable. From the very beginning, the computer industry became very adept at having a "standard" that fell short of what most understand a standard to be. The three most common "tricks" are the following:

1. Write the standard so general that almost anything can be made compliant by just using the right wording; that is, all a supplier has to do is to get a good wordsmith. The first SP88 standard focused on terminology. However, it is the underlying concepts that are really important. The NAMUR document focused on the concepts and had the greater impact. The suppliers got the message—produce this and we will buy it; otherwise, we will not.

2. If there are multiple ways of doing something, incorporate all of them into the standard. The ISA had to take this approach to get a standard for fieldbus. However, just because a technology is written into a standard does not guarantee its survival. Market forces will create winners and losers.

3. The standard basically says that "if a feature is implemented, then you must do it this way." However, the suppliers have the option of not implementing the feature at all and can still otherwise claim compliance to the standard.

From the outset, the computer industry has been extremely dynamic. In such an environment, standards are not viewed in a positive light. We must find a better way to do whatever, and when we find it, we do not want any standard to obstruct our introduction of this into the market. The success of the computer industry as compared with others (including the process industries) suggests that its approach has merits.

Standards in the computer industry do not drive the technology; they follow. Just look at the evolution of Ethernet. By the time a standard appeared, Ethernet was clearly the winner over token-ring and other competitors. The market chose the winning technology, and then a standard followed.

Within the processes industries, the prospect of an occasional "collision" generated responses akin to cardiac arrest, with some declaring that they would never install Ethernet in control applications. Technically, Ethernet transmissions are not deterministic. Collisions are possible even with 100 Mb or 1000 Mb Ethernets, but with no consequences to an application. With its widespread use, hardware and software for Ethernet is readily available and very economical. Consequently, Ethernet is routinely installed in the process industries, even in control applications. With computer technology, it is best to go with the flow—swimming upstream is futile.

6.1. STRUCTURED PROGRAMMING

The bane of all software development is "spaghetti code." Containing numerous paths with considerable redundant code, the result is excessively complex, making subsequent support extremely challenging. The basic objective of structured programming is to avoid such code. Certain programming constructs were introduced, and if properly applied, would avoid at least the worst of spaghetti code. The **GOTO** statement of FORTRAN was demonized, and it does indeed appear throughout spaghetti code.

Structured programming led to the development of languages such as C (which ironically provides a **GOTO** statement as a standard feature of the language). Features derived from structured programming were eventually added to languages such as FORTRAN and Basic. But in practice, good programmers were using structured programming and object-oriented concepts all along, despite not being supported by features of the language. Good organization of the program logic is the key, regardless of the terms used to describe it. In conventional programming, this often led to the separation of the executable code from the data needed for the code to perform its task. In essence, this became the basis for object-oriented programming, which led to the enhancement of C to C++.

6.1.1. Table-Driven Software

Process control systems have long employed a variation of object oriented programming known as "table-driven software." Control functions such as proportional-integral-derivative (PID) were implemented by creating a table organized as follows:

1. A column of the table was dedicated to each parameter required to execute a PID control computation. Examples included an index or pointer to the source of the process variable (PV), the option specified for direct/reverse action, the controller gain, and so on.
2. A row of the table was reserved for each PID loop to be configured.

In effect, each row of the table constituted a block of data equivalent to an object in the context of object-oriented programming.

A single set of executable code processed all loops. To execute a loop, this code was activated and provided either an index for the row in the table or a pointer to the block of data for the PID loop to be executed. With this approach, PID loops are configured, not programmed. Configuring a loop means defining a row in the table or building a data block, and has evolved from processing a set of fixed-format records to interactive screens to create or modify a PID loop. However, this approach has some limitations:

1. The number of rows in the table was fixed at the outset, which imposed an upper limit on the number of PID loops that could be configured.
2. Each row of the table contained the same number of columns. When the table is used only for PID loops, this is not a problem. But if the table is extended to provide different types of control blocks (one being the PID), a fixed number of columns means that the same amount of storage is reserved for each block. Problems arise when implementing control blocks such as moving average and dead time that require locations to store past values of the block input.

True object-oriented implementations create the object at the time a PID loop or control block is configured. This permits as many objects to be created as required, and each object can be created with the amount of storage required for that control block.

6.1.2. Structured Logic for Batch

For automating a flexible batch process a structured approach is essential. The key to structuring the logic is to separate the following:

Logic pertaining to the production equipment. This logic is usually specific to a site, and could be very different from one site to another. For example, one plant might transfer material A through a header from a storage tank whereas another plant charges material A from drums. Each is capable of charging material A to the vessel, but it is done in very different manners.

Logic pertaining to the product being manufactured. This logic should not be site-specific. That is, the same product-specific logic for manufacturing hexamethylchickenwire should be used in all plants that produce this product.

The ideal is to keep these entirely separate.

This can be illustrated with material transfers. Suppose the manufacture of hexamethylchickenwire requires that specified amounts of materials A, B, and C be charged to a vessel. Logic must be provided to execute each material transfer, the possibilities including the following:

1. An automatic transfer via a header with block valves
2. Charged from drums placed on a platform scale
3. Manually weighed and emptied into the vessel.

The logic for each transfer is entirely dependent on the equipment used to execute the transfer. In no way is it dependent on the product being manufactured. For manufacturing hexamethylchickenwire, the transfers are performed in exactly the same manner as manufacturing any other product.

The amount of each material to be transferred certainly depends on the product being manufactured. However, the order may also depend on the product. For hexamethylchickenwire the sequence is to add A, then B, and finally C. But for another product, the order could be exactly the reverse, that is, C, then B, and finally A.

Unless there is some characteristic of the equipment that restricts the order to A, then B, and finally C, the sequence of the charging must not be part of the logic related to the process equipment. In most cases, the equipment places no restrictions on the sequence of charging the materials, so this sequence must be part of the logic pertaining to the product being manufactured.

Formula and procedure. The desire is for specifications such as the order of charging materials to be incorporated into the product recipe. A limited number of such options can be provided through the formula. In practice, this approach works reasonably well for single-product batch facilities, but not for flexible batch facilities. If you ask a chemist to write the recipe for hexamethylchickenwire, the result would be something akin to the following:

Formula:
  X kg of A
  Y kg of B
  Z kg of C
Processing instructions:
  1. Charge A.
  2. Charge B.
  3. Charge C.

In flexible batch facilities, the recipe encompasses all of this. In practice, the desire is for all product-specific logic to be specified as part of the recipe.

The previously mentioned recipe contains no information pertaining to the equipment. The exact mechanism for charging A cannot be determined from the recipe. From the perspective of the products, how A is charged is immaterial. Provided a mechanism is available for transferring X kg of A to the vessel, the plant can manufacture hexamethylchickenwire.

6.1.3. Obstacles

Difficulties can arise when separating the logic pertaining to the equipment from the logic pertaining to the product. Consider the following situation:

1. Materials A and C are both charged through the same header.
2. After transferring material C, the transfer piping must be flushed with material A.

The customary approach is to short the initial transfer of A by the amount of A required to flush the transfer piping. The amount of A required to flush the piping depends on the equipment, specifically, the length and size of the piping in the transfer header.

Issues such as the amounts required for flushing could be added to the product recipe, but if so, the resulting recipe is site-specific. That is, the flush could be provided as follows:

Formula:

X kg of A (total)
Y kg of B
Z kg of C
F kg of A (for flush)

Processing instructions:

1. Charge A, less the amount required to flush the header.
2. Charge B.
3. Charge C, followed by flushing the header with A.

The value of F is specific to a site, which makes the recipe site-specific. The preferable approach is to keep such values out of the recipe and handle them separately.

The flush amount should be specified as part of the equipment-specific logic, preferably in a single location where its value can be easily changed. What if the material used for the flush depends on the product being manufactured? The flush amount can be specified in kilograms of water, which can then be multiplied by the specific gravity of the flush material to obtain the quantity of flush material required.

When incorporated into each recipe, changing the flush amount necessitates a change in every recipe in which material C is charged. Such a change is easy, but those with experience know what will happen. The number of recipes in a flexible batch plant can easily be a hundred or more. This gives opportunities for errors, such as the following:

1. One (or possibly more) of the recipes will not be modified.
2. An incorrect value will be specified in one (or possibly more) of the recipes.

And in plants where version control is maintained for the recipes, this is a change that must be duly recorded and documented for each affected recipe.

Implementing the logic. Consider providing separate functions or subroutines for charging each raw material. Suppose the following routines are available:

> **CHARGE_A.** Charge a specified amount of raw material A to the vessel, with the amount to be charged specified via a parameter.
>
> **CHARGE_B.** Charge a specified amount of raw material B to the vessel, with the amount to be charged specified via a parameter.
>
> **CHARGE_C.** Charge a specified amount of raw material C to the vessel, with the amount to be charged specified via a parameter.

The logic within each routine would be specific to a given facility. However, it would be independent of the product being manufactured.

Ignoring the issues pertaining to the flush following the transfer of material C, the charging of the raw materials for hexamethylchickenwire could be accomplished by the following segment of code:

CALL CHARGE_A(AMOUNT_A)
CALL CHARGE_B(AMOUNT_B)
CALL CHARGE_C(AMOUNT_C),

where **AMOUNT_A, AMOUNT_B**, and **AMOUNT_C** are obtained from the formula within the product recipe. This logic is entirely dependent on the product being manufactured and should be the same for every production site.

But as noted previously, some recipes require that the materials be charged in the reverse order. For this, the segment of code should be as follows:

CALL CHARGE_C(AMOUNT_C)
CALL CHARGE_B(AMOUNT_B)
CALL CHARGE_A(AMOUNT_A).

The routines being called are exactly the same, but the order in which they are called depends on the product recipe.

This requirement can be met via the following approaches:

1. For each product recipe, implement the sequence in code that is specific to that product recipe. In other words, a program must be written and maintained for each product recipe. This is unacceptable. Except for values in the formula, changes to the product recipe must be implemented by changing the program. Furthermore, the program must be relinked should any change be made to **CHARGE_A, CHARGE_B**, and so on.

2. Use dynamic linking to routines such as **CHARGE_A, CHARGE_B**, and so on. For C++, such a capability is supported in the form of

dynamic link libraries (DLLs). However, this requires a level of software sophistication not commonly found in the process industries.

Developers of process control systems designed for automating batch processes have devised various approaches to obtain a result comparable to the latter, but there is no universally accepted solution.

6.2. PRODUCT RECIPES AND PRODUCT BATCHES

The general organization of the logic for automating a batch process presented in Figure 6.1 reflects a structured approach. The division between product-specific logic and equipment-specific logic occurs at the phases.

6.2.1. Product Recipe

In the structure illustrated in Figure 6.1, everything from the phases up is product-specific and should be part of the product recipe. As the product recipe preferably encompasses all of the product-specific logic, the contents of the product recipe should be organized as follows:

Formula and procedure. The product recipe consists of a formula and a procedure. The formula is a data set, and is the counterpart to the list of ingredients for a recipe in a cookbook. The procedure encompasses all of the instructions for manufacturing the batch. However, the formula does not necessarily contain all parameter values. Some may appear within the instructions contained within the procedure.

Operations. The procedure is divided into operations. When multiple items of equipment (such as a reactor and three feed tanks) are required to

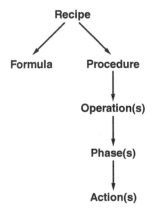

Figure 6.1. Structure for the logic required to automate a batch process.

manufacture the product batch, one approach is to define an operation that specifies the processing to be performed on each item of equipment, that is, an operation for the reactor, an operation for feed tank A, an operation for feed tank B, and an operation for feed tank C. Although this basis for dividing the procedure into operations is perhaps the easiest to envision, other bases are possible.

Phases. An operation is divided into phases. Basically, an operation is a list of phases to be activated in a specified sequence. As this is specific to a product, the product recipe must specify the phases to be activated and the order in which they are activated.

The logic that constitutes a phase specifies the specific actions to be performed to accomplish the objective of the phase. These actions are specific to the equipment being used to manufacture the batch. Consequently, the actions that comprise a phase should not be part of the product recipe.

6.2.2. Product Batch

The product recipe and product batch are defined as follows:

Product recipe. It is basically the master plan for manufacturing a batch of product.

Product batch. It is a unit of production in a batch process. In industries like specialty chemicals, the quantity of product that results from a product batch is fixed. In such industries, it is not possible to manufacture a half-batch.

For each product batch, a product recipe must be designated.

The use of product recipes for consecutive product batches can be as follows:

Same product recipe is used for a number of consecutive product batches. In such facilities, scheduling is often in terms of the desired amount of product to be manufactured. This amount is divided by the standard yield for each product batch to determine the number of product batches required to produce the desired amount of product.

Product recipe differs from batch to batch. Occasionally, the same product recipe is used for two or possibly three consecutive product batches, but this is not the norm. Scheduling is in terms of batches, with the standard yield used to determine the expected production amount.

6.2.3. Standard Batch Size

The origin of most product recipes is a development group that uses small-scale equipment (relative to the production facilities) to produce sufficient quantity of a product for field trials. Scaling up to the production equipment means adjusting quantities such as raw material amounts for the equipment capacities. However, parameters such as temperatures must not be scaled. The capacity of the production equipment very likely differs from one production site to another, and sometimes, different equipment sizes are available within the same production site (e.g., a 5000-L reactor and a 20,000-L reactor are available).

The recipe is often expressed in terms of a reference quantity, the most common being the following:

Equipment capacity. The recipe may be expressed for a vessel with a standard volume in liters, such as a 1000-L vessel.

Product amount. The recipe may be expressed such that the expected result is a standard amount of product, such as 1000 kg (a metric ton).

For each product recipe, the expected yield is established from production history for that product. The product amount and the equipment capacity are related by a factor whose units are the product yield in kilogram per liter of vessel capacity.

6.2.4. Product Batch ID

Each product batch is assigned an identifier that herein will be referred to as the batch ID. Although the batch ID is entirely numeric in the occasional facility, the batch ID could potentially be any string of printable characters. Most plants have rules for constructing the batch ID from parameters such as calendar date, equipment tags, and so on. Sometimes, corporate standards are applied, but often the construction of the batch ID is at the discretion of the plant.

One purpose of the batch ID is for tracking quality information. Some facilities perform tests on every batch of product; others perform tests more selectively. For example, the product batch might be assumed to meet specifications if certain criteria are met (charges are within tolerances, temperatures are attained and maintained within tolerances, etc.). Any item of data that contributed in any way to the assessment of the quality of the product batch must be retained, preferably in a manner that the data can be quickly retrieved.

Another requirement that is becoming more common is material tracking. Most raw materials are provided in lots with associated identifiers. Material tracking involves capturing the actual amount and lot identifier for each raw material consumed in making the product batch. Typical objectives for material tracking include the following:

1. To better respond to quality complaints from customers
2. To limit the scope of any recall that may be required.

6.3. FORMULA

The major division of a product recipe is between the formula and the procedure (or processing instructions). In older systems, the formula was often a list of numerical values, some of which could be used to designate flags (yes/no) or numerical options. But as systems have evolved, most systems now permit the formula to contain text strings.

An example where text strings are very useful is for the logic to charge material from drums via a platform scale. The operators manually position the drum on the platform scale, insert the transfer piping, and so on. Any material that is supplied as a liquid in a drum can be charged. The logic for the material transfer is the same, regardless of the contents of the drum.

6.3.1. Designating Materials

The amount to be transferred is certainly one of the parameters supplied to the logic that executes the material transfer. Another should be an identifier for the material to be transferred. The options for this identifier are the following:

Chemical name of the material. These are not always meaningful to operators.

Common name of the material. This must be sufficiently explicit so that there is no confusion as to what is to be charged.

Product code for the material. These explicitly designate the material to be charged in a language-independent manner (an advantage in a multinational organization). However, product codes are not instantly recognized by humans. But by linking the Material Safety Data Sheet (MSDS) to the product code, the name and all other relevant information is available to the process operator.

This may also apply to intermediates. Materials produced within the process are sometimes stored in drums for subsequent use in other production operations.

6.3.2. Specifying Values

Traditionally product chemists write recipes in a manner surprisingly similar to how recipes appear in a cookbook. Consider the recipe for Mom's Pecan Pie presented in Table 1.1. In this recipe, parameters are supplied in the following two ways:

Within the formula. Only the raw material amounts are specified in the formula, which is appropriately designated the "List of Ingredients."

Within the directions (processing instructions or procedure). In step 6, the baking time and temperature are specified within the statement "Bake 60 minutes at 350°F."

Within the chemical industry, recipes are prepared by product chemists. The closer the control system recipe matches the recipe as written by the product chemists, the better.

6.4. OPERATIONS

A distinguishing characteristic of batch processing is that activities can be performed in parallel.

The recipe for Mom's Pecan Pie in Table 1.1 will be used as an example where operations can be defined. Just remember that the production facility is the kitchen in one's home; commercial bakeries are very different. The steps in the directions in Table 1.1 appear to be performed in a sequential manner. However, this is not actually the case.

6.4.1. Parallel Operations within a Product Batch

Step 6 for this recipe specifies "Bake 60 minutes at 175°C (350°F)." This recipe assumes that the cook will preheat the oven to 175°C (350°F), although some recipes include statements such as "Preheat oven to 175°C (350°F)." In either case, the cook should not wait until arriving at step 6 to turn on the oven. But where in the sequence of the steps in Table 1.1 should the cook turn on the oven? The conservative approach would be to turn on the oven at the start of step 1, but the preferable approach would be to turn the oven on at a time that the temperature attains 175°C (350°F) just as step 5 is completed. The point is that things happen in parallel—the oven is heating while the steps in Table 1.1 are being executed.

Another example of parallel activities is a recipe for a cake that requires icing. Steps similar to those in Table 1.1 specify how the batter is to be prepared and the cake is to be baked. Separate steps are specified for preparing the icing. Most cooks perform these steps while the cake is baking in the oven. These are clearly parallel activities.

A very similar situation occurs in some chemical batches. Prior to initiating a reaction, two tasks must be completed:

1. A solution must be prepared in a feed tank.
2. The reactor must be prepped by charging a heel, starting the agitator, and adjusting the temperature of the heel.

These activities should definitely proceed in parallel. The one requiring the longest time is preferably initiated at the first opportunity. The other must be initiated so that it completes in a timely manner.

At this point the focus is on the requirements for parallel activities within the logic pertaining to the product, that is, within the processing instructions or procedure section of the product recipe. The need for parallel activities can also arise within the logic pertaining to the equipment. This will be addressed in the subsequent chapter on sequence logic.

Again using the recipe for Mom's Pecan Pie in Table 1.1 as the basis for this discussion, the activities can be divided as follows:

1. Blending the ingredients
2. Baking the pie, including preheating the oven.

Table 6.1 is the result of rewriting the recipe in Table 1.1 in terms of two operations, one named **Mixing** for blending the ingredients and the other named **Baking** for all activities pertaining to baking the pie. Steps have been added to both operations for transferring the pie from the mixer to the oven.

Coordination. Even though operations can proceed in parallel, they are not entirely independent. For the recipe in Table 6.1, step 2 of operation **Baking** cannot commence until operation **Mixing** has completed. The two operations are interconnected in two ways:

1. Step 6 of operation **Blending** cannot proceed until the oven has been preheated in accordance with step 1 of operation **Baking**.

TABLE 6.1. Operations within a Recipe

Mom's Pecan Pie

List of Ingredients

4 eggs	200 gm (1 cup) sugar
240 mL (1 cup) light Karo	4 gm (½ tbsp) flour
1½ gm (¼ tsp) salt	5 mL (1 tsp) vanilla
60 mL (¼ cup) butter	220 gm (2 cups) pecans

Directions

Operation **Mixing**
1. Beat eggs until foamy.
2. Add sugar, Karo, flour, salt, and vanilla.
3. Beat well.
4. Stir in melted butter and pecans.
5. Pour into pie crust.
6. Transfer pie to oven.

Operation **Baking**
1. Preheat oven to 175°C (350°F).
2. Transfer pie from mixer.
3. Bake 60 minutes at 175°C (350°F).

2. Step 3 of operation **Baking** cannot commence until the pie has been transferred to the oven in accordance with step 6 of operation **Blending**.

Basically, the activities in the two operations must be synchronized at step "Transfer pie to oven" of operation **Mixing** and step "Transfer pie from mixer" of operation **Baking**.

To produce Mom's Pecan Pie requires two items of equipment: a mixer and an oven. The operations in Table 6.1 express the activities that must occur on each item of equipment. Recipes in the chemical industry frequently do the same. While the main focus is on equipment such as reactors, the product recipe normally encompasses associated equipment installed for purposes such as the following:

1. A feed is a mixture whose preparation must be completed before the reaction can commence. The proper amounts of certain raw materials must be mixed and possibly heated or cooled to a specified temperature.
2. The product from the reactor must be further processed, examples being filtering, pH adjustment, decanting, and so on.

The preference is to incorporate all required activities into a single product recipe. Providing separate recipes for such activities creates opportunities for mistakes.

Preferably, a single recipe covers all activities between one set of storage equipment (vessels, drums, pallets, etc.) for the feeds to another set of storage equipment for the products. The feeds may be raw materials or intermediates (material produced by other production activities within the plant or possibly a sister plant). The products may be salable products, intermediates, by-products, wastes, and so on.

6.4.2. Parallel Product Batches

Operations can be performed in parallel either for a given product batch or for other product batches. Using Mom's Pecan Pie as the example, operations are performed in parallel for the following situations:

1. For a given pie, preheating the oven occurs in parallel with mixing the ingredients.
2. When producing two consecutive pies, the ingredients for the second pie can be mixed while the first pie is baking.

For the latter case, the two pies do not have to be the same type.

Figure 6.2 illustrates baking a pecan pie immediately followed by baking a cherry pie. The pecan pie is baking while the ingredients for the cherry pie are being mixed. In effect, an "on-the-fly" transition is made from baking the

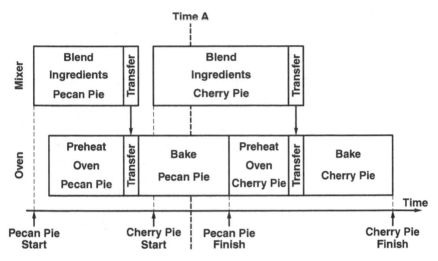

Figure 6.2. Execution of product batches.

pecan pie to baking the cherry pie. Even though the two batches utilize a different product recipe, the process is not stopped between the two batches.

6.4.3. Transitions from One Product to Another

The following view of a chemical production process is rarely appropriate:

1. Prepare the process for making one product. This could involve activities such as switching the contents of a storage tank from one raw material to another.
2. Execute a specified number of batches of this product.
3. Clean the process equipment or do whatever else is required to return the process equipment to its initial state.

Instead of stopping the process for a transition, the norm is to make "on-the-fly" product transitions. The objective is to minimize production downtime. Sometimes this requires careful scheduling. For example, it may be possible to make on-the-fly transitions from **Product A** to **Product B** and then to **Product C**. But before returning to **Product A**, the process must be stopped and appropriately prepared.

Especially in the programmable logic controller (PLC) world, the initial concept for automating a batch process was to prepare separate programs for manufacturing each product. You want to make pecan pies? Load the pecan pie program and make pecan pies. You want to make cherry pies? Load the cherry pie program and make cherry pies. In the 1970 time frame, the major market for PLCs was the automotive industry. For production processes such as stamping machines, the machine would be configured for stamping the front-left fender for a specific car model, the front-left fenders program loaded into the controller, and the specified number of fenders would be produced. Time to make rear-left fenders? Reconfigure the machine, load the rear-left fenders program, and produce the specified number of rear-left fenders. Today, "flexible" has entered the manufacturing industries as well, but this is the way it worked in the 1970s.

But batch chemical plants were making "on-the-fly" transitions in the 1970s. The concept of a program for each product had a fundamental flaw. Consider "Time A" in Figure 6.2. A pecan pie is baking; a cherry pie is mixing. What program is loaded in the controller? To use specific programs for each product, the process must be shut down to transition from one product to another. For most batch chemical processes, this results in an excessive loss of production time, making "on-the-fly" transitions (at least between most products) the general practice within the industry.

6.5. PHASES

Software designers are thoroughly schooled on the use of functions or subroutines in the construction of a large program. Advantages include the following:

1. Ideally eliminate, but in practice greatly reduce, redundant code.
2. Reuse code from other projects.
3. More efficient code development and testing. Each routine can be assigned to an individual or team to develop and test, thereby permitting them to work independently of others. When all is completed, the results are integrated to produce the final product.

In the automation of batch processes, the counterpart to a function or subroutine is a **phase**. However, there is an additional advantage, namely, the separation of equipment-dependent logic from the product-dependent logic.

6.5.1. Definition of Phases

Using a prior example, the addition of materials A, B, and C for manufacturing hexamethylchickenwire could be performed by the following segment of code:

CALL CHARGE_A(AMOUNT_A)
CALL CHARGE_B(AMOUNT_B)
CALL CHARGE_C(AMOUNT_C).

In procedural languages, **CHARGE_A**, **CHARGE_B**, and **CHARGE_C** would be functions or subroutines. But in the automation of batch processes, they would be phases.

The phases must be formulated in accordance with the following objectives:

1. No product-dependent logic is incorporated into the phase.
2. The phase consists entirely of equipment-dependent logic.

The logic within the phase is definitely dependent on the plant equipment and consequently is almost always site-dependent.

6.5.2. Phase Parameters

When activated, the phase must be supplied with values for all relevant parameters. In the previous example, the only parameter is the amount to be charged. When a given product is to be manufactured at several different sites within the company, the following stipulations should be imposed:

1. The function provided by the phase should be the same. A logical extension is that its name should be the same.
2. The list of parameters supplied when activating the phase should be the same.

This permits a product recipe currently in use at one of the sites to be copied to another site with a minimum of changes. In the ideal world, there would be no changes, but actual implementations rarely achieve this ideal.

6.5.3. List of Available Phases

One of the tasks in designing the approach to automate a batch facility is to develop a list or menu of phases, which entails defining the purpose of each phase, specifying the list of parameters, and so on. The two extreme philosophies for undertaking this task are the following:

From the equipment perspective. Knowing the details of the equipment within the process, one can define all of the process functions that the equipment can perform. If one develops a phase for each of these, then anything which the equipment is capable of performing is available within the batch automation framework. The downside of this is that some phases may never be used, so the work involved in implementing those phases is wasted.

From the product perspective. From an analysis of the recipes currently in use at a site, a list of process functions that are actually being used can be prepared. In a flexible batch facility with a hundred recipes or more, time constraints usually prohibit the analysis to encompass all recipes. Instead, the analysis is based on the dozen or so recipes that are used most frequently. The resulting list will very likely be missing a few phases.

As the remaining recipes are commissioned, an occasional phase will not be available, which means someone must create the phase, usually on short notice. The frequency of such occurrences should reduce with time, but may never drop to zero. New product recipes are commissioned from time to time, so the possibility of a needed phase not being available remains.

These are the extremes. The typical approach is to start with the latter, and then review the resulting list of phases with people experienced with the products being manufactured to hopefully catch at least the most obvious omissions.

6.5.4. Simple versus Complex Phases

When formulating the list of phases, another consideration is which of the following two approaches should be followed:

Long list of simple phases. Each individual phase is formulated to accomplish a very specific objective. The logic within the phase is simple, with fewer parameters required when activating the phase. But to cover all required functions, the list of phases can be quite lengthy.

Short list of complex phases. Each individual phase is formulated to accomplish a number of related objectives. The logic within the phase is more complex, and more parameters are required when activating the phase. However, the list of phases is much shorter.

Which is preferable? Opinions are voiced, sometimes very loudly. But in the end, they are just opinions.

The **Mixing** operation for Mom's Pecan Pie in Table 6.1 will be used to illustrate the issues. Table 6.2 illustrates the two options. In the approach in Table 6.2a, simple phases are provided for the following:

1. Each solid to be added
2. Mixing.

This requires 12 different phases. The approach in Table 6.2b relies on one phase that can add any solid and another phase that can mix to whatever degree is required. This approach reduces the required number of phases to five, but two of these phases are more complex, each now requiring a parameter. In real life, the number of parameters for the complex phases can become substantial, such as 12 or more. The increased complexity of the phase logic is comparable.

Suppose the need arises to make a version of a pecan pie that contains raisins. With the first approach, a simple phase is required to add raisins. If no

TABLE 6.2. Phases within the Operation Mixing

(a) Long List of Simple Phases

Operation Mixing	Phase	Parameters
1. Beat eggs until foamy.	**AddEggs**	4
	MixFoamy	
2. Add sugar, Karo, flour, salt, and vanilla.	**AddSugar**	200 gm (1 cup)
	AddFlour	4 gm (½ tbsp)
	AddSalt	1½ gm (¼ tsp)
	AddVanilla	5 mL (1 tsp)
3. Beat well.	**MixWell**	
4. Stir in melted butter and pecans.	**AddButter**	60 mL (¼ cup)
	AddPecans	2 cups
	Stir	
5. Pour into pie crust.	**PourPieCrust**	
6. Transfer pie to oven.	**TransferOven**	

(b) Short List of Complex Phases

Operation Mixing	Phase	Parameters
1. Beat eggs until foamy.	**AddEggs**	4
	Mix	Foamy
2. Add sugar, Karo, flour, salt, and vanilla.	**AddSolid**	200 gm (1 cup), "Sugar"
	AddSolid	4 gm (½ tbsp), "Flour"
	AddSolid	1½ gm (¼ tsp), "Salt"
	AddSolid	5 mL (1 tsp), "Vanilla"
3. Beat well.	**Mix**	Well
4. Stir in melted butter and pecans.	**AddSolid**	60 mL (¼ cup), "Butter"
	AddSolid	220 gm (2 cups), "Pecans"
	Mix	Stir
5. Pour into pie crust.	**PourPieCrust**	
6. Transfer pie to oven.	**TransferOven**	

such phase exists, one must be written. With the second approach, the complex phase is hopefully written to be able to add whatever solid the recipe requires. If so, the capability to add raisins already exists.

6.6. ACTIONS

To accomplish its purpose, a phase must perform an appropriate sequence of actions that depend on the plant equipment. Consequently, the logic within a phase is usually site-dependent. However, the logic should not contain anything that is dependent on the product being manufactured.

6.6.1. Logic for a Phase

To illustrate, both operations in Table 6.2 begin with a phase **AddEggs** that must be supplied one parameter when activated—the number of eggs to be added. The value of the parameter depends on the product being manufactured, the value normally being specified within the product recipe. However, the logic for adding eggs is not product-dependent. A typical approach is expressed as follows:

For each egg, do as follows:
- Pick up egg gently.
- Carefully crack egg shell.
- Dump contents into bowl.
- Discard egg shell.

The cook does it this way for a pecan pie, for a chocolate cake, or any other product that requires adding all of the egg. Of course, some recipes require only the egg yolk, and some require only the egg white. These require a slightly different approach. Either separate phases for **AddEggs**, **AddYolksOnly**, and **AddWhitesOnly** are provided, or a more complex phase with an additional parameter for **All**, **Yolks**, or **Whites** can be developed.

Control logic can be broadly classified as follows:

Discrete logic. Most interlocks are implemented using discrete logic. For example, the fill valve to a vessel must be closed if the high-level switch indicates that liquid is present. The relay ladder logic used to program PLCs is very efficient at rapidly executing discrete logic.

Continuous control. In the process industries, the workhorse is the PID control function. Most modern digital control systems provide a number of additional continuous control functions. Figure 5.5 and Figure 5.6 implement dribble flow using continuous control functions.

Sequence control. Automating batch processes often requires that actions be performed in sequence. The previous example of adding eggs is sequence logic—do this, then do that, and so on. Simple sequences can be implemented in either relay ladder diagrams or continuous control functions, but more complex sequences require different approaches. A subsequent chapter is devoted to this subject.

6.6.2. Sequence Logic

Sequence control is an alternative to the continuous control functions in Figure 5.5 and Figure 5.6 for implementing dribble flow. Although not included in Figure 5.5 and Figure 5.6, the following logic encompasses the block valves for starting and stopping the transfer:

1. Start with all block valves closed and the dribble flow valve(s) closed.
2. Reset the flow totalizer.
3. Compute the pre-preset as the amount to be transferred (the preset) less the total quantity to be added via dribble flow.
4. Open the block valves for transferring the desired material.
5. Open the dribble flow valve(s) for full flow.
6. Wait until the flow total attains the pre-preset.
7. Set the dribble flow valve(s) for dribble flow.
8. Wait until the flow total attains the preset.
9. Close the dribble flow valve(s).
10. Close all block valves.

This accomplishes everything with sequence logic. An alternative is a combination of sequence control and continuous control, as illustrated by the following sequence:

1. Start with all block valves closed.
2. Reset the flow totalizer.
3. Compute the pre-preset as the amount to be transferred (the preset) less the total quantity to be added via dribble flow.
4. Write the values for the pre-preset and preset to either the two comparators in Figure 5.5 or the cutoff block in Figure 5.6.
5. Open the block valves for transferring the desired material.
6. Wait until the flow total attains the preset.
7. Close the dribble flow valve(s).
8. Close all block valves.

In older models of digital control systems, the continuous control functions were executed more efficiently and more rapidly than sequence control, giving the latter approach an advantage. However, modern digital control systems execute sequence logic quite rapidly as well. The approach is often determined by personal preference.

One issue is missing from the previously mentioned sequence logic. What if the block valve for a different raw material opens while the transfer is in progress? The raw material transfer must be stopped as rapidly as possible, usually by closing all block valves. The sequence logic presented earlier can be thought of as the "normal logic"—what should be done if everything is proceeding as expected. Additional logic, sometimes referred to as the "what if" or "failure" logic, is required to react to abnormal situations. This will also be explored in the subsequent chapter on sequence logic.

REFERENCES

1. Batch Control Part 1: Models and Terminology, American National Standards Institute, ANSI/ISA-88.00.01-2010.
2. Anforderungen an Systeme zur Rezeptfahrweise, NAMUR, NE033 (March 31, 1993).

7

BATCH UNIT OR PROCESS UNIT

The logic within a phase is specific to the plant equipment on which the actions are to be performed. One option is to permit the logic to have access to any process point. With this approach, the logic can do whatever is required. Although acceptable in a small application, issues arise in large applications. Having access to all points creates an undesirable possibility: an error in the logic within a phase could affect a part of the process other than the part to which the phase pertains. Such errors in the logic are not reproducible, making them very difficult to find.

Such considerations give rise to the concept of a batch unit or process unit. These define the part or subset of the facility to which a phase pertains. Basically, a phase is written specifically for a batch unit. Each operation within the product recipe is activated to execute on a batch unit, and each phase within the operation executes on that batch unit. By restricting access to this part of the process, the possibility that errors in the logic can affect other parts of the process is greatly reduced if not eliminated.

Like dividing a large program into subroutines, there is no hard and fast approach to dividing a process facility into batch units. In most cases, a batch unit corresponds to a unit operation such as a reactor, a separation column, and so on. However, there are exceptions.

When an operation is restricted to execute only on a designated batch unit, one statement can be made with regard to defining batch units:

Control of Batch Processes, First Edition. Cecil L. Smith.
© 2014 John Wiley & Sons, Inc. Published 2014 by John Wiley & Sons, Inc.

If two (or more) operations are to execute simultaneously, separate batch units must be defined for each operation.

That is, batch units must be defined such that any requirements for parallel activities within the product-specific logic (as specified in the product recipe) can be met.

7.1. DEFINING A BATCH UNIT

Batch units are normally assigned a tag name or other suitable identifier. Although implementations vary considerably, the configuration parameters for a batch unit include parameters such as type of batch unit (use will be explained later in this chapter), a parameter that reflects the capacity of the equipment (e.g., volume of vessel in liters), and others. However, the most important component is the list of I/O points associated with the plant equipment for which the batch unit is defined.

Most control systems designed for automating batch processes provide facilities to make defining and modifying batch units quite easy. While advantageous, this only goes so far. The definition of phases very much depends on how the batch units are defined. Once an application is well underway, redefining the batch units also means that the definitions of the phases must be modified accordingly. As a result, one tends to get locked into the batch unit definitions early in a project, so it is prudent to take the time to get them right at the outset.

7.1.1. Reactor with Three Feed Tanks

Figure 7.1 illustrates a reactor with three feed tanks. Manufacturing a batch of product typically proceeds as follows:

1. Solutions are prepared in each of the three feed tanks.
2. A heel is added to the reactor, the agitator is started, and the temperature adjusted to the desired value.
3. The solutions from the three feed tanks are fed in the proper ratios.
4. After the co-feeds complete, the temperature in the reactor is increased and held for a specified time.
5. The reactor contents are cooled and discharged.

At step 4 in the above-mentioned sequence, the feed tanks are no longer in use for the batch currently in the reactor. This permits solutions to be prepared for the next batch in parallel with step 4 in the above-mentioned sequence for the current batch.

For product batches that require that three solutions be prepared, the product recipe must contain at least four operations:

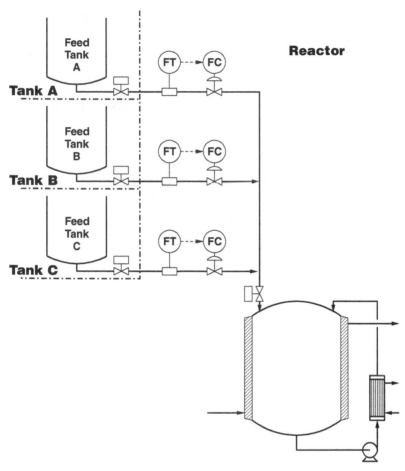

Figure 7.1. Reactor with three feed tanks.

1. Operation for preparing the solution in **Tank A**
2. Operation for preparing the solution in **Tank B**
3. Operation for preparing the solution in **Tank C**
4. Operation for the **Reactor**.

The block valve installed on the discharge of each feed tank should be open only when the co-feed is in progress. Otherwise, the block valve should be closed. When closed, the feed tank is isolated from the reactor (and the other feed tanks); consequently, events occurring in the feed tank have no effect on the reactor (or any other feed tank). This permits the following:

1. All solutions can be prepared simultaneously.
2. Preparation of the solutions for the next batch can be undertaken while the previous batch is still in progress on the **Reactor**.

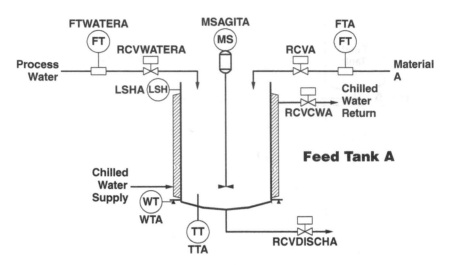

Figure 7.2. I/O points for Tank A.

I/O list for feed tank. Figure 7.2 shows **Tank A** in greater detail. The solution is prepared by mixing water, material A, and small amounts of additional materials that are added manually by the process operator. After thoroughly mixing, the solution is cooled using chilled water. Heat transfer is better in metallic feed tanks than in glass-lined reactors, so chilling the feeds reduces the amount of heat that must be removed from the reactor.

The I/O points consist of the following:

Inputs:

TTA	**Tank A** temperature
WTA	**Tank A** weight
FTWATERA	Flow of water to **Tank A**
FTA	Flow of material A to **Tank A**
LSHA	High-level switch for **Tank A**

Outputs:

RCVWATERA	Block valve for charging water to **Tank A**
RCVA	Block valve for charging material A to **Tank A**
RCVCWA	Block valve for chilled water to **Tank A**
RCVDISCHA	Block valve for discharge from Tank A to reactor
MSAGITA	**Tank A** agitator motor start/stop

Sometimes the inputs and outputs are further subdivided into analog and discrete. The discrete points in the I/O list fall into two categories:

1. Those whose states are determined by the logic executing on the batch unit. These are discrete outputs that drive field devices.

2. Those whose state is available to the logic but not determined by the logic. All discrete inputs fall into this category. However, this category also includes discrete outputs whose state is determined by logic executing on another batch unit.

Most implementations also permit the list to include other types of points, specifically, continuous control blocks such as proportional-integral-derivative (PID) control. By restricting access to only those points in the I/O list, a phase executing on a batch unit can change set points, switch modes, and so on, of only those PID control blocks in its I/O list.

The sequence logic that comprises a phase is supplied the I/O list for the batch unit on which the phase is executing. By restricting access to only the points in the I/O list, errors in the phase logic should not lead to problems in other parts of the process. From this perspective, the limitation needs only to be applied to the outputs. Input points can be freely accessed with no concern for ramifications beyond the batch unit. But as will be explained shortly, there are other reasons for accessing all points (inputs and outputs) through the I/O list associated with the batch unit.

7.1.2. Complex Batch Units

For batch units defined for process equipment such as reactors, the I/O list can become lengthy, possibly including more than a hundred points. The implications for complexity deserve serious consideration. The options are analogous to those for defining phases:

1. Define the minimum number of batch units, even if for some the I/O list is very long.
2. Define a larger number of batch units, the objective being to avoid very long I/O lists.

The downside of the latter approach is that more coordination is required between the activities proceeding on the respective batch units.

Using the reactor in Figure 7.1 as the example, several I/O points are associated with the co-feeds (control valves, PID controllers, ratio blocks, etc.). These are only relevant while the co-feed is in progress, that is, when the block valve that admits the co-feeds into the reactor is open. However, the co-feeds are specific to the reactor and would never be used in a manner independent of the reactor. The incentive to define a separate batch unit for the co-feeds is to reduce the length of the I/O list for the reactor. The co-feed phase for the reactor cannot directly execute the co-feeds. To proceed with the co-feeds, the phase must coordinate with a separate phase executing on the batch unit defined for the co-feeds.

Figure 7.3 illustrates separating two items of equipment from the reactor batch unit.

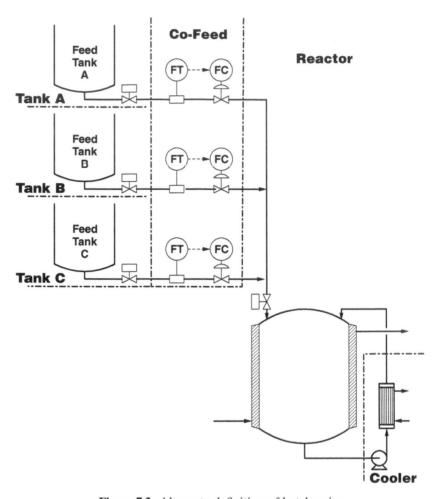

Figure 7.3. Alternate definition of batch units.

Co-feeds. Chapter 3 discussed the control issues for a similar arrangement.
Pump-around cooler. Although not shown in Figure 7.3, block valves are
  usually provided to block off the cooler when not in use.

Usually, such separate batch units make sense only if the resulting I/O list for
the reactor is significantly shorter.

7.2. SUPPORTING EQUIPMENT

To keep the diagrams as simple as possible, only the key I/O points are
included in Figure 7.1 and Figure 7.3. The jacket is another source of a large

number of I/O points, especially for jacket switching arrangements such as in Figure 3.14. As compared with the previous discussion pertaining to a separate batch unit for the co-feeds, the situation with regard to the jacket is different in two respects:

1. The reactor can be isolated from the co-feeds by closing the block valve that admits the co-feed to the reactor. There is no counterpart for the jacket.
2. The reactor can never be operated independently of its jacket.

7.2.1. Switched Jacket

With the switched jacket arrangement illustrated in Figure 3.14, four possible operating modes are possible:

1. Heating with steam
2. Cooling with tower water
3. Cooling with chilled water
4. Cooling with glycol.

This leads to a potential complication for any phase that charges a material to the reactor. To cover all possibilities, the phase must provide for the following:

1. Charge the material while heating with steam.
2. Charge the material while cooling with tower water.
3. Charge the material while cooling with chilled water.
4. Charge the material while cooling with glycol.

Incorporating these options into the phase for charging a material increases the complexity of the phase.

These issues potentially apply to all of the phases. The result is a significant increase in the complexity of each phase, with the additional logic being substantially the same for all phases. This raises the usual issues pertaining to redundant code—any modification in the logic must be implemented in all of the phases.

7.2.2. Pressure Control

For some reactors, the issues increase rapidly. Many reactors are equipped with various alternatives for pressure control, such as the following:

Vacuum. To avoid the possibility of cross-contamination through a shared vacuum system, the trend is to provide dedicated vacuum equipment for each reactor. The associated logic must provide for starting and stopping the vacuum equipment. Especially when dual-stage vacuum pumps are installed, this involves more than a simple on/off switch, although usually much simpler than switching a jacket from one heating/cooling mode to another.

Atmospheric. The pressure is maintained either slightly below atmospheric pressure (to avoid leaks to the outside world) or slightly above atmospheric pressure (to avoid oxygen leaking into the reactor). This requires a source of an inert gas that can be admitted to the reactor when the pressure drops and a vent system to which gas can be released when the pressure rises.

Pressure. When one of the reactants is a gas, provision must be made for the reactor to be pressurized with this gas. While the reaction is in progress, pressure can be controlled by increasing or decreasing the flow rate of the gaseous reactant. Venting the gas is a loss of raw material, but may be required at the end of the reaction phase.

Some phases must be capable of coping with all three. Using a phase for charging a material as the example, the following capabilities are required:

1. Charge material while maintaining vacuum.
2. Charge material under atmospheric pressure control.
3. Charge material under pressure control.

Temperature control is essential for all reacting systems. Consequently, these pressure control options must be combined with the four heating/cooling modes described earlier. The results are the following:

1. Charge material while maintaining vacuum and heating with steam.
2. Charge material while maintaining vacuum and cooling with tower water.
3. And so on.

With four cooling modes and three pressure control modes, the possible combinations total 4 × 3 = 12.

7.2.3. Separation Column

Should a separation column and condenser be installed on the reactor, the situation becomes even more extreme. Such columns potentially provide the following three modes of operation:

Total reflux. The column and condenser are essentially operated so as to remove heat from the reacting medium.

Product draw. Sometimes this is a salable product. Sometimes the material withdrawn appears on the product side of an equilibrium reaction, so removal of the product shifts the equilibrium toward the product side.

Oil/water separator. Usually one phase is withdrawn and the other returned to the column.

With three modes of operation for the separation column, the possible combinations total $4 \times 3 \times 3 = 36$.

7.2.4. Implications for Phase Logic

For phases that execute tasks such as charging a raw material, the logic within the phase clearly has to address all issues associated with transferring the material—resetting totalizers, opening block valves to start the transfer, closing block valves upon conclusion of the transfer, determining the total amount transferred, and verifying that the actual amount is within a specified tolerance of the target amount. However, control issues pertaining to jackets, pressure control equipment, and so on, can be implemented separately. Prior to the start of a material transfer, the jacket should be switched to the proper mode, the pressure control must be operating in the desired manner, and so on.

For most material transfers, the same conditions are maintained throughout the material transfer. But should a phase require that something be changed at some point during the execution of the phase, the capability must be provided for directing the jacket logic, the pressure control logic, and so on, to implement the change.

Using the jacket as the example, the product recipe must specify the heating mode and the target for the reactor temperature, but the logic pertaining to the jacket should do the rest. For a switched jacket, the logic should maintain the current status of the jacket, but probably with an additional operating mode designated "All Off" or perhaps "Jacket Empty." The possibilities for transferring the parameters pertaining to the jacket logic from the product recipe are as follows:

1. A mechanism is provided to transfer the parameters directly from the product recipe to the jacket logic.
2. The parameters are first provided to a phase (such as a material transfer phase), which merely transfers them to the jacket logic. Most phases would do this at the very start of the phase, but a few will need to make subsequent adjustments.

For a switched jacket, the product recipe specifies the heating mode along with values such as the target for the temperature. Suppose the product recipe specifies that the contents of the reactor be heated with steam to a temperature of 90°C. The following logic is required:

1. The jacket control logic determines if a switch in the operating mode is required and if so executes the switch. This requires sequence logic, albeit of the repetitive type.
2. Once the jacket is switched to the appropriate operating mode, adjusting the temperature is normally accomplished with continuous control logic. The jacket logic switches the appropriate controller(s) to automatic and specifies their set points.

7.2.5. Between Product Batches

The phases that comprise an operation can only be active when a batch is in progress. Some systems permit a phase to be directly activated by a process operator, and if so, executes independently of a product batch. Logic such as the jacket switching logic must definitely function while a product batch is active, with parameters such as the operating mode and targets originating from the product recipe. The possibilities for the transition from one product batch to another include the following:

1. Terminate the jacket logic at the end of each batch. When the jacket logic is activated for the next batch, the operating mode is determined from the current states of the various block valves. However, the operating mode may not be entirely clear. For example, rarely is a direct indication of jacket empty or full provided, and if full, whether its contents are water or glycol.
2. When the jacket logic is terminated, always switch the jacket to a pre-determined operating mode, such as All Off or Jacket Empty. When activated for the next batch, the jacket logic should start by verifying that the states of the various valves are consistent with this operating mode.
3. Keep the jacket logic active between product batches. This permits the jacket logic to always know the current operating mode. It may also be useful to provide a mechanism whereby the process operator can instruct the jacket logic to change the operating mode.

Of these, keeping the jacket logic active at all times is usually preferable. Since no phase is active between product batches, the logic must be implemented separately from the main phase logic. One approach is to implement sequence logic that is dedicated to jacket control, with certain parameters (temperature target, heating/cooling mode, etc.) from the product recipe transferred in some manner to this dedicated sequence logic. However, this sequence logic is not implemented as a phase within the structured batch logic.

This creates the potential for a downside to this approach. Many batch facilities contain multiple items of identical (or very similar) equipment for which batch units are configured. As will be explained in a subsequent section of this chapter, most process controls provide the capability for the user to develop a phase for one batch unit, but then execute that same phase on other "identical" batch units. Obviously, the objective is to avoid redundant code. However, this capability applies to phases. Does it also apply to sequences that are dedicated to items of equipment? That is, the dedicated sequences are not executed as phases, so this capability does not automatically extend to the dedicated sequences.

Modern control systems, including programmable logic controllers (PLCs), support functions, subroutines, or something comparable. These can potentially be used to write one sequence that can be executed on different batch units, that is, on different I/O points. However, limitations are possible. Numerical values, text strings, and so on, can surely be passed as parameters. However, can a point address be passed as a parameter? As always, the devil is in the details.

7.3. STEP PROGRAMMER

Execution of sequence logic on a batch unit almost always involves specifying the states of the field devices associated with that batch unit. Consider the martini mixer in Figure 7.4. When using a structured approach to batch logic, a batch unit would be configured for the martini mixer.

The mixer is used to prepare several martinis that are then consumed at the discretion of customers. For the preparation of each martini, the process proceeds by a repetitive sequence of the following steps:

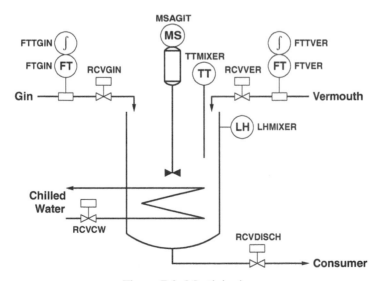

Figure 7.4. Martini mixer.

TABLE 7.1. Process States for Normal Logic

Step	RCVDISC	RCVGIN	RCVVER	MSAGIT	RCVCW
Add gin	Closed	Open	Closed	Off	Closed
Add vermouth	Closed	Closed	Open	Off	Closed
Mix	Closed	Closed	Closed	On	Closed
Chill	Closed	Closed	Closed	On	Open
Consume	Don't care	Closed	Closed	Don't care	Don't care
Shutdown	Closed	Closed	Closed	Off	Closed

Step 1. Close all valves (shutdown).

Step 2. Add gin.

Step 3. Add vermouth.

Step 4. Mix.

Step 5. Chill.

Step 6. Consume (adding olives and the like are at the discretion of the consumer).

Although commonly encountered within the context of batch processing, the term "step" is unfortunately used in a variety of ways. As used herein, "step" will designate a progression of the sequence logic.

7.3.1. Field Device States

As the process proceeds from one step to another, the discrete field devices must operate in accordance with the device states indicated in Table 7.1. For each step, the desired state of every field device is specified. For two-state field devices, there are three possible specifications for the state:

1. State 0 (closed or off)
2. State 1 (open or on)
3. Don't care, meaning that the state of the device is at the discretion of the process operator, the sequence logic running on another batch unit, or whatever.

For step **Consume**, the discharge valve state, the agitator state, and the chilled water valve state are **Don't Care**. The discharge valve is opened at the discretion of each customer to fill his or her martini glass. Once the level in the mixer drops to the point that the agitator blades are not submerged, someone should stop the agitator and close the chilled water valve. Alternatively, one could automate this as follows:

1. Install a level switch that permits the agitator to run only if the level is above the location of the switch.

2. Add logic that permits the chilled water valve to be open only if the
   agitator is running.

These are process interlocks and would be implemented within the process
controls.

Proceeding from one step to the next normally involves operating more
than one device. For example, proceeding from **AddGin** to **AddVermouth**
requires closing the gin valve and opening the vermouth valve. Two approaches
are possible:

1. Operate each device individually. To be sure that all devices are in the
   proper states, the logic must output to all devices that are required to be
   in a specified state, not only those whose state changes.
2. For each step in the sequence, specify the required state of each device.
   A facility is required that, given a value for the step, drives each device
   to its specified state.

As the number of devices increases, the latter approach becomes more
attractive.

7.3.2. Drum Timer

In conventionally controlled plants, some repetitive sequences were automated by a hardware device referred to as a "drum timer." The older washing
machines used a residential version of such a device to control the wash
sequence. These timers sequenced the various two-state devices in accordance
with the configuration of the drum timer. The normal use was to start these
timers at the first step and allow them to proceed from step to step based solely
on the elapsed time. For each step for the drum timer, the configuration parameters consisted of the following:

1. State for each controlled device
2. Elapsed time for the step.

The user could start the sequence at a step other than the first one, but few
other options were provided.

Drum timers were used to automate repetitive mixing sequences in applications such as the martini mixer. To add the desired amount of gin, the gin valve
was opened for a specified amount of time. The time for this step was "tuned"
in the field, often by someone sampling the product and conducting tests. For
the martini mixer, this likely consisted only of a taste test. If the result was
"needs more gin," then increase the time for the **AddGin** step. Rather crude
and not very precise, but as long as the martinis taste good, who cares. Unfortunately, modern industrial processes are not so forgiving, so timing each step
is not normally adequate.

The original drum timers were mechanical. A motor rotated the drum at a fixed speed. In completely mechanical versions, pegs on the surface of the drum tripped switches to open/close valves and to start/stop motors. In electronic versions, conductive strips on the surface of the drum provided continuity in the circuit for opening a valve or starting a motor. Implementing the counterpart in software provides more flexibility and makes parameters such as the time for each step much easier to change.

7.3.3. Enhancements to Drum Timer

Especially when implemented in software, a number of enhancements are possible, one being to provide options other than time to initiate the advance from one step to the next. Using the martini mixer as the example, the basis for exiting each step and proceeding to the next should be as follows:

Shutdown. Operator action to start the preparation of martinis. The sequence begins with **AddGin**.

AddGin. The proper amount of gin has been added (based on the flow totalizer).

AddVermouth. The proper amount of vermouth has been added (based on the flow totalizer).

Mix. The mixer has run for a specified time.

Chill. The mixture attains the desired temperature.

Consume. Operator action indicating that the vessel is empty. The logic proceeds to **Shutdown**, from which the logic can be restarted at **AddGin**.

With such enhancements, the term "drum timer" is no longer appropriate. Furthermore, implementing in software eliminates the physical drum or its equivalent. Herein the software implementation will be referred to as a "step programmer."

With a software implementation, a number of other extensions are possible, including the following:

1. Provide more options for proceeding from one step to another, such as the ability to skip steps, to jump steps (normally proceed to the next step, but under certain conditions proceed to another step), to repeat steps (such as a loop), and so on.
2. Monitor discrete inputs, discrete outputs, and alarms for detection of abnormal conditions within the process that prevent the normal sequence logic from proceeding.
3. Extend the configuration parameters for the step to include "analog" values such as PID set points, ratio coefficients, and so on.

7.3.4. I/O List

Configuring a step programmer requires an I/O list similar to the I/O list for a batch unit. This creates two possibilities:

1. Associate each step programmer with a batch unit, with the step programmer using the I/O configured for the batch unit.
2. Independently specify the I/O list for each step programmer.

As more functions are incorporated into the step programmer, its I/O list tends to approach that of the batch unit.

7.4. FAILURE CONSIDERATIONS

Continuing to use the martini mixer in Figure 7.4 as the basis for the discussion, the logic presented so far pertains to what must occur under the expected or "normal" conditions. Even in the best plants, things occasionally stray from normal.

Before starting another mix, the process operator should verify that the mixer is empty. With the instrumentation illustrated in Figure 7.4, there is no independent means to confirm that the vessel is indeed empty. What happens if the operator mistakenly starts another mix with the vessel half full? The vessel would overflow unless appropriate actions are taken.

All issues addressed in this chapter apply to process interlocks, as defined in the introductory chapter. Consequently, the logic is usually implemented within the process controls. Implementing external to the process controls is an option, but as will be discussed shortly, there are consequences that must be addressed. Anything pertaining to personnel safety is implemented in equipment dedicated to that function.

7.4.1. Indications of a Problem

The high-level switch illustrated in Figure 7.4 is installed specifically so that actions can be taken that prevent vessel overflows. Should the level in the vessel actuate the high-level switch, the feeds to the vessel must be stopped. However, there is no need to dump the contents of the vessel, and the agitator can continue to run.

The possibilities for indications that the process is not operating in the expected manner include the following:

1. A discrete input, such as the high level switch in Figure 7.4, indicates that a problem exists.
2. A discrete field device controlled by logic executing on another batch unit is not in the appropriate state.

3. The state feedback(s) for one (or more) of the discrete field devices controlled by the logic executing on the batch unit are not consistent with the values of the output to that field device.

Such events are referred to as failures, requiring appropriate actions to respond to the problem.

7.4.2. Shutdown State(s)

Upon detection of a failure, the normal progression of the sequence executing on a batch unit must be suspended in a fashion that the process is placed in a state that permits operations personnel to ascertain the root cause of the problem, determine what options are available, and to choose one.

In this context, "shutdown" means that the normal progression of the batch logic has been suspended in an appropriate manner. It could mean total shutdown in which the contents of vessels are discharged, resulting in a complete loss of the batch. But if possible, the desire is to devise a plan to recover from the failure and continue with the production of the batch. Even producing a batch of lower quality product is usually preferable to dumping vessels and incurring costs to dispose of the contents.

What is possible depends on the nature of the process. For reactors, is it possible to stop all feeds but to maintain mixing, temperatures, and so on? For columns, it is possible to switch to total reflux or to recycle all product streams to the feed tank?

Two possible approaches are available for defining the shutdown states:

1. Define the appropriate shutdown state for each possible failure condition.
2. Based on process considerations, define the possible shutdown states, and then for each failure condition, determine which of these shutdown states is appropriate.

In most cases, the second approach is preferable. One consideration in defining shutdown states is that each must be clearly explained to the process operators. Most modern process controls permit a large number of shutdown states to be defined, but far more than most process operators can remember.

Restricting the number of shutdown states to a small number may mean that a slightly different shutdown state would be appropriate for one or more failure conditions, but the benefits are rarely worth the effort. Failures should be infrequent; if not, this is a problem that needs attention.

7.4.3. Bypassing or Overriding Process Interlocks

This discussion applies only to process interlocks; bypassing or overriding a safety interlock, if permitted at all, must be severely restricted.

For the martini mixer in Figure 7.4, suppose the high level switch fails in the actuated state, indicating an abnormally high level. There are two options:

1. Suspend production operations until the faulty level switch is repaired. But especially when this occurs in the wee hours of the morning, the result could be several hours of lost production. Needless to say, the plant manager will not be amused.

2. First, ascertain that the level is not high, and furthermore, is approximately correct for this stage of the mix. If a process interlocks, bypassing the logic associated with the high-level condition is acceptable, which at least permits the mix currently in progress to be completed. An added advantage is that maintenance personnel can make the repair on an empty vessel instead of one partially filled with chemicals.

Bypassing or overriding a process interlock is usually restricted (such as to the shift supervisor), but the expectation is that occasionally bypassing this logic is appropriate. Most process controls make provisions for this to be done, the following being two examples:

1. As discussed in a previous chapter, most discrete device drivers permit a faulty state feedback to be ignored.

2. For a discrete input, most provide some feature that permits a specified value to be substituted for the current input value. One approach is to place the discrete point "off-scan" and substitute the desired value. PLCs provide the equivalent capability by permitting the value of a discrete input to be "forced" to the desired value.

Preferably, the capability to bypass or override a process interlock is available to operations personnel (such as the shift supervisor), and can be done without involving maintenance personnel.

7.4.4. Implementation of Process Interlocks

Herein the following three approaches will be examined for implementing process interlocks in the process controls:

1. Discrete logic only
2. Discrete logic coupled with a discrete device driver
3. Step programmer.

Figure 7.5 illustrates a header that can feed water to one of four vessels: **Feed Tank A**, **Feed Tank B**, **Feed Tank C**, and **Reactor**. The flow measurement for water is shared between the three feed tanks and the reactor. A batch unit will be configured for each of the four vessels. Each valve is controlled by the batch unit with which it is associated.

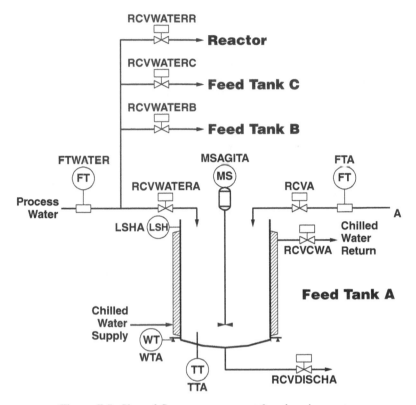

Figure 7.5. Shared flow measurement for charging water.

Although the logic executing on batch unit **Tank A** does not control the valves for admitting water to **Feed Tank B**, **Feed Tank C**, or **Reactor**, these valves must be closed when water is flowing to **Feed Tank A**. Similar considerations apply to all vessels, but herein attention is directed only to the logic for **Feed Tank A**. An interlock can be viewed from two perspectives:

1. A list of all conditions that must exist for the water valve for **Feed Tank A** to open. The valve is permitted to open only if the logical **AND** of all of these conditions is **TRUE**.

2. A list of all conditions for which the water valve for **Feed Tank A** must be closed should any one of the conditions exist. The valve must be closed if the logical **OR** of all of these conditions is **TRUE**.

The discussion that follows is based on the first approach. In practice, either can be used.

The water valve for **Feed Tank A** is permitted to open only if all of the following conditions are **TRUE**:

1. High level switch in **Feed Tank A** is not actuated. The discrete input to the process controls senses the state of the normally closed switch. Consequently, input state 0 indicates high level; input state 1 indicates normal (not high) level.

2. Water valve to **Feed Tank B** is closed as indicated by a limit switch on the closed position. An input state of 1 confirms that the valve is closed.

3. Water valve to **Feed Tank C** is closed as indicated by a limit switch on the closed position. An input state of 1 confirms that the valve is closed.

4. Water valve to **Reactor** is closed as indicated by a limit switch on the closed position. An input state of 1 confirms that the valve is closed.

7.4.5. Discrete Logic Only

The first issue to address is the commanded state for the **Feed Tank A** water valve. The physical push buttons formerly installed in panels have been replaced with simulated push buttons on a graphic display. By pressing the appropriate push button, the operator can command the valve to open or close. Discrete variable **FeedTankAWaterValveCommandedState** will be used for the commanded state from the process operator.

When no process interlock is required, the physical output to the valve is always the same as the commanded state. When required, the logic for the process interlock is inserted between the commanded state and the physical output to the valve.

For the **Feed Tank A** water valve, the logic consists of the logical **AND** of the following conditions:

FeedTankAWaterValveCommandedState. The commanded state for the valve. Logic state 1 commands valve to open; logic state 0 commands valve to close.

FeedTankAWaterHighLevelSwitch. **Feed Tank A** high-level switch. Logic state 1 is level normal (not high); logic state 0 is level high.

FeedTankBWaterValveLimitSwitchStateClosed. Limit switch for state closed for **Feed Tank B** water valve. Logic state 1 confirms that valve is closed.

FeedTankCWaterValveLimitSwitchStateClosed. Limit switch for state closed for **Feed Tank C** water valve. Logic state 1 confirms that valve is closed.

ReactorWaterValveLimitSwitchStateClosed. Limit switch for state closed for **Reactor** water valve. Logic state 1 confirms that valve is closed.

This logic is expressed by the following logic statement:

FeedTankAWaterValvePhysicalOutput =
FeedTankAWaterValveCommandedState &
FeedTankAWaterHighLevelSwitch &

FeedTankBWaterValveLimitSwitchStateClosed &
FeedTankCWaterValveLimitSwitchStateClosed &
ReactorWaterValveLimitSwitchStateClosed

Following the C notation, the logical operators used herein are as follows:

|—logical **OR**
&—logical **AND**
!—logical **NOT**.

The above-mentioned logic relies solely on the limit switch to confirm that a valve is closed. Although implementing the limit switches as proximity switches within the valve actuator has significantly reduced the frequency, limit switch failures continue to be a possibility. One could argue that to accept that a valve is closed, both of the following conditions must be met:

1. The physical output to the valve must correspond to valve closed (logic state 0).
2. The limit switch for state closed must confirm that the valve is closed (logic state 1).

In hardwired implementations, this additional logic is not normally included, mainly because of the increased complexity and the need for additional hardware. But when implemented as software in process controls, no additional hardware is required—all is in software and the added logic can be easily implemented, incurring "no additional cost." Unfortunately, the latter is not quite right. Nothing is free! Costs are always associated with additional complexity. Do the benefits (overflowing the vessel is less likely) justify the cost associated with the extra complexity? Unfortunately, neither the benefits nor the costs are easy to quantify in monetary units.

In all software implementations of process interlocks, "creeping elegance" is an issue. At the design stage, the usual practice is to pose "what ifs." Each can be addressed, but at the expense of additional complexity. In many projects, the organization that poses the "what ifs" will get paid to add the logic. Although the potential exists for a conflict of interest, the actual concern of the designers is that consequences will arise due to a condition that was not identified.

Overly complex logic seems to always complicate plant start-up. Will a valve with 50 conditions that must be met to open actually open? The likely answer is "not too often, and not for too long." One suspects that many of these conditions are not really necessary. Now for the irony. Once the unneeded conditions have been identified, the organization that was paid to add the logic will likely be paid to remove it.

7.4.6. Discrete Device Driver Coupled with Discrete Logic

For the process water header in Figure 7.5, suppose a discrete device driver is configured for all valves. A valve is now commanded to open or close through the discrete device driver.

The approach used to implement process interlocks depends on whether or not the discrete device driver provides a logic input (0 or 1; **TRUE** or **FALSE**) that supports one of the following features (or something equivalent):

OK to Open. A logic value of 1 on this input permits the valve to open.

Force to Safe State. A logic value of 1 on this input forces the valve to close.

Force Tracking. A logic value of 1 on this input places the discrete device driver in the tracking mode. When tracking is active, a second input specifies the state for the output of the discrete device driver.

As these are equivalent, **OK to Open** will be used in the following examples.

To implement the process interlock, the conditions that must be met are incorporated into a logic expression that is **TRUE** when the valve is permitted to open. For the **Feed Tank B** water valve, the logic expression can use any of the following for confirming that the valve is in the closed state:

1. Limit switch on the closed position must indicate that the valve is closed.
2. Limit switch on the closed position must indicate that the valve is closed and the physical output to the valve must be driving the valve to the closed state.
3. The discrete device driver for the valve must indicate that the valve is in the closed state. The states indicated by the discrete device driver are **Open, Closed, Transition**, and **Invalid**. The **Closed** state is indicated only if the following are true:
   a. The physical output to the valve is driving the valve to the closed state.
   b. The limit switch on the closed position indicates that the valve is closed.

The **Feed Tank C** water valve and the **Reactor** water valve would be treated similarly.

When a discrete device driver is available, the logic expression for process interlocks should use the state indicated by the discrete device driver. All available information on the state of the valve is incorporated into the logic within the discrete device driver. In addition, a faulty limit switch can be easily ignored using the features provided by the discrete device driver.

This logic expression is used in both of the approaches illustrated in Figure 7.6 that incorporate the process interlock with the discrete device driver:

"OK to Open" (or its equivalent) is supported. As illustrated in Figure 7.6a, the result of the logic statement for the process interlock is used for

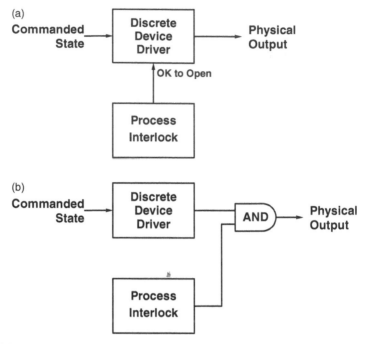

Figure 7.6. Implementing process interlocks: (a) through the discrete device driver and (b) between discrete device driver and physical output.

the "OK to Open" input to the discrete device driver. The output of the discrete device driver is the physical output to the valve.

"OK to Open" (or its equivalent) is not supported. As illustrated in Figure 7.6b, the output of the discrete device driver is the commanded state for the valve. The physical output to the valve is the logical **AND** of the following:

1. The output of the discrete device driver (the commanded state for the valve)
2. The result of the logic expression that expresses the conditions for the process interlock. This logic expression is **TRUE** when the valve is permitted to open.

Either approach enforces the process interlock, but there is a very important difference. Suppose the commanded state for the valve is **Open**, but the process interlock prevents the valve from opening. The difference is as follows:

Figure 7.6a. The process interlock is enforced by the discrete device driver. The state of the valve is **Closed** (the physical output to the valve is driving the valve to the closed state, and the limit switch on the closed position confirms that the valve is closed. The discrete device driver will correctly indicate the state of the valve as **Closed**. But in addition, the discrete device driver can also indicate that the valve is "Interlocked," "Tracking," or something equivalent. In this way, the process operator knows that something within the process controls is preventing the valve from opening.

Figure 7.6b. The discrete device driver is provided no information regarding the process interlock. Suppose the following are true:

1. The valve is commanded to open through the discrete device driver. The output of the discrete device driver will be the same as the commanded state, that is, valve open.

2. The process interlock is preventing the valve from opening. Even when the valve is commanded to open, the physical output to the valve is driving the valve to the closed state. The valve is in fact closed, and the input from the limit switch confirms that the valve is closed.

The output from the discrete device driver corresponds to the valve open state, but the limit switch indicates that the valve is closed. After the specified transition time has elapsed, the discrete device driver indicates that the state of the valve is **Invalid**. The **Invalid** state is normally understood to indicate a problem within the valve itself. But in this case, the valve is functioning properly; the **Invalid** state is the result of the process interlock.

The arrangement in Figure 7.6a is preferred, but can only be used if the discrete device driver supports the required features.

7.4.7. Step Programmer

Process interlocks can also be enforced utilizing a step programmer with the following specifications for each step:

1. States for all discrete outputs controlled by the step programmer, with one option for the state being "Don't Care"

2. Desired states for monitored discrete points (this includes discrete inputs, alarms, and discrete outputs that are not controlled by this step programmer).

When directed to a specified step, the step programmer does the following:

1. At the start of the step, the programmer drives the controlled discrete outputs to their specified states.

2. As long as this step is active, the programmer monitors the following:

   (a) The controlled discrete outputs for remaining in their specified states

   (b) The monitored discrete points for being in their specified states.

   Should any discrepancy between actual state and specified state occur, the step programmer proceeds to a step that results in a process shutdown of some type.

Monitoring the valves in Figure 7.5 can be by either of the following:

1. Monitor the limit switch that provides the state feedback for the **Closed** state.
2. Monitor the state indicated by the discrete device driver for the respective valve. If the valve is to be **Closed**, a failure is initiated should the state indicated by the discrete device driver be **Open** or **Invalid**. Failures are not usually initiated by the **Transition** state (a temporary state).

A distinct advantage of the latter approach is that an "Ignore" for a state feedback is entirely handled by the discrete device driver.

Utilizing a step programmer associated with the batch unit for **Feed Tank A** offers two advantages:

1. The discrete logic expression to provide the input for **OK to Open** is not required.
2. When the step programmer has commanded **Feed Tank A** water valve to open, opening any other water valve will cause the **Feed Tank A** water valve to close, and execution of the sequence logic will be suspended.

For the batch unit for **Feed Tank A**, the configuration of the step programmer for the **Feed Water** step would be as follows:

	Device	State
Controlled outputs:	**Feed Tank A** water valve	Open
Monitored outputs:	**Feed Tank B** water valve	Closed
	Feed Tank C water valve	Closed
	Reactor water valve	Closed

For all other steps, the configuration would be configured as follows:

	Device	State
Controlled outputs:	**Feed Tank A** water valve	Closed
Monitored outputs:	**Feed Tank B** water valve	Don't Care
	Feed Tank C water valve	Don't Care
	Reactor water valve	Don't Care

With these specifications, the other water valves are monitored for the closed state only while the **Feed Tank A** water valve is open (during the **Feed Water** step). At other times, the state of the other water valves are of no concern, so opening one of these valves has no effect on the sequence logic executing on **Feed Tank A**.

7.4.8. Implications for Sequence Logic

On an indication of high level in the martini mixer in Figure 7.4, each of the approaches described previously prevents the vessel from overflowing. However, an additional issue must be addressed: execution of the sequence logic must be suspended.

When the sequence logic is directing the step programmer from one step to another, provision is usually made for a failure detected by the step programmer to also suspend execution of the sequence logic. In many cases, directing the step programmer to a shutdown step is sufficient to place the process in an appropriate shutdown state. But in some cases, the sequence logic also needs to take actions; that is, a short sequence is also required to attain an appropriate shutdown state.

When discrete logic is used in lieu of a step programmer, execution of the sequence logic must also be suspended. If the **AddGin** and **AddVermouth** steps are timed, the sequence logic would proceed to the **Mixing, Chill,** and **Consume** steps. Customers are unlikely to find the product acceptable. If these steps are terminated on the flow total attaining the target value, the sequence logic would not proceed. The sequence logic continues to be executed, but the condition to proceed to the next step will never be satisfied. Presumably the process operator would eventually notice that the mix is taking too much time and would investigate. However, it is preferable to suspend execution of the sequence logic and clearly so indicate to the operator.

Upon suspension of the sequence logic associated with a shutdown initiated by a process interlock, an informational message that conveys what caused the shutdown to be initiated is certainly appropriate. But should this event be an alarm that must be acknowledged by the process operator? An alarm is usually configured to alert the process operator to the high level condition within the mixer. This raises the potential for two alarms in rapid succession—a liquid level high alarm followed by an alarm on the suspension of the sequence logic.

A good definition of a redundant alarm is one that informs the process operator of something that the operator already knows. Does the operator know that the liquid level high alarm causes execution of the sequence logic to be suspended? There are three possible answers:

Yes. The alarm on suspending the sequence logic is redundant.

No. The alarm on suspending the sequence logic is appropriate.

Some do and some do not. Issuing an alarm on suspending the sequence logic can be viewed as the conservative approach. But there is a counterargument. Always taking a conservative approach in a batch process can lead to an excessive number of alarms, which impairs the performance of the process operators—instead of doing something useful, the operator is preoccupied with acknowledging alarms.

In addition to the fuzzy answer as to what the operator already knows, there is another complication. While feeding water to **Feed Tank A**, suppose the water valve to **Feed Tank B** opens. Why would this occur? There are various possibilities, but only one will be described. Mechanisms will be presented shortly so that the sequence logic executing on **Feed Tank B** will wait until the water has been transferred to **Feed Tank A**. However, plants fear that production is stopped because the logic within the process controls is preventing a valve from opening. To address this fear, they demand that a mechanism be available to bypass the logic and open the valve. Such mechanisms may not be properly utilized.

Two possible approaches can alert the operator that a shutdown has occurred:

1. Configure an alarm for the water valve for **Feed Tank B** being open (or not closed). Process control systems provide mechanisms to enable (arm) an alarm under certain conditions, but disable (disarm) that alarm under other conditions. For this alarm, these features must be properly used.
2. Issue an alarm that the sequence logic executing on **Feed Tank A** has been placed in shutdown, the reason being that the water valve to **Feed Tank B** is not closed.

The latter approach is probably the simplest, but one issue is raised. This alarm is certainly appropriate if the shutdown is initiated because the water valve for **Feed Tank B**, **Feed Tank C**, or the **Reactor** opens. But if the shutdown is the result of vessel high level (for which an alarm is issued), the result could be viewed as a redundant alarm.

Process controls provide a variety of mechanisms to avoid redundant alarms, including the following:

1. Features to arm and disarm alarms
2. Alarm filters that are designed to suppress the redundant alarms
3. Intelligent alarm packages.

All have a common requirement—logic must be developed to provide the basis for avoiding the redundant alarms.

Developing this logic is not a trivial undertaking. Several individuals must be involved to carefully formulate and extensively review the logic, making the effort time-consuming and costly. Even so, the possibility exists that some alarm will be incorrectly deemed to be redundant and consequently suppressed, which means that the process operator is not properly informed of some issue within the process.

Although suppressing all redundant alarms is unlikely to be achieved, the most obnoxious ones can be suppressed, and the tools to do so are readily available. But the most common obstacle is that the plant manager who rails about the excessive number of alarms in a batch facility will refuse to pay for the effort to correct the problem.

7.4.9. Failures Initiated by Momentary Events

For the martini mixer in Figure 7.4, the following is a possible occurrence. The agitator motor is equipped with an overload switch that opens should the motor draw more power than its design allows. Normally, the overload switch is incorporated into the motor starter circuitry that is often external to the process controls. The resulting implementation is discrete logic that stops the agitator on a motor overload.

The state feedback from the motor starter circuitry provides the indication to the process controls that the motor has stopped. The discrete device driver detects that the state feedback is not consistent with the desired state of the agitator, and sets the state of the device as **Invalid**. Basically, the output to the agitator motor should cause it to be running, but the state feedback indicates that the agitator is stopped.

The appropriate response is to place the martini preparation logic in the **Shutdown** state. Several approaches could be used to accomplish this, but all have a characteristic that deserves consideration. The behavior would be as follows:

1. Motor is running.
2. Overload occurs, and overload switch opens.
3. Motor stops.
4. Overload switch closes (a stopped motor is not in overload).
5. Motor remains stopped. The motor starter circuitry is designed such that the motor does not automatically restart when the overload switch closes.

Basically, the overload switch opens for only a very short period of time. Once power is removed from the motor, the overload switch quickly closes.

The following possibility applies to any process switch (level, pressure, temperature, etc.) providing an input to logic that behaves in a similar manner:

1. The process is operating normally.
2. A process switch detects a condition that causes the logic to place the process in a shutdown state.
3. The condition that caused the shutdown to be initiated inherently clears with the process in the shutdown state.

Maintenance personnel are now faced with a serious question—what initiated the shutdown?

Using the agitator motor as the example, how does maintenance know that the shutdown was initiated because of a motor overload condition? With no indication, the likely approach is to start the motor and see if the shutdown occurs again. The shortcomings to this approach include the following:

1. There could be some consequences for the plant equipment.
2. The condition that caused the shutdown may not occur instantly. Sometimes it takes an hour or more. Some shutdowns are associated with transients in the process that occur only occasionally.

So far this discussion has pertained to shutdowns that are properly initiated. However, the possibility exists that the condition causing the failure was falsely indicated, possibility due to an intermittent fault within the equipment detecting the condition.

Two possibilities for providing an indication of the condition that initiated the shutdown are as follows:

1. The logic should issue a clear message indicating the condition that initiated the shutdown.
2. Logic should be added to "latch" the condition that initiated the shutdown. The logic for latching could be provided externally or within the process controls. The latter is usually advantageous because provision must also be made for clearing the latch, the possibilities being the following:
   (a) Automatically clear the latch when the process is restarted.
   (b) If the latch is set, require that the latch be manually reset before the process can be restarted.

Arguments can be made that only maintenance personnel need access to the latched information. However, most process operations personnel also want access to this information.

7.5. COORDINATION

One of the considerations when defining batch units is the capability to execute an operation on a batch unit with no consideration as to what is occurring on other batch units. However, there are times that the activities on two or more batch units must be coordinated.

This need arises for the reaction system in Figure 7.1. Consider **Feed Tank A** and the **Reactor**. The following activities can proceed in parallel:

1. Preparing the solution in **Feed Tank A**.
2. Preparing the **Reactor**, including charging the heel, starting the agitator, and so on.

But when it is time to transfer the solution from **Feed Tank A** to the **Reactor**, activities in the two batch units must be coordinated.

The transfer must not commence until the following two events have occurred:

1. Preparation of the solution in **Feed Tank A** has completed
2. The **Reactor** is ready to receive the solution. The heel must be charged, the agitator must be started, and so on.

One of these will usually be completed prior to the other. Preferably, the norm is for the preparation of the solution to be completed before the **Reactor** is ready to receive it. The **Reactor** is the most expensive part of the equipment, and idle time in the **Reactor** means lost production.

But occasionally, the **Reactor** will be ready to receive the solution before its preparation is complete in **Feed Tank A**. Usually this is the result of unexpected events such as equipment malfunctions. But since these will occasionally occur, the coordination logic must cope with both of the following:

1. Preparation of the solution completes before the **Reactor** is ready to receive it.
2. The **Reactor** is ready to receive the solution before its preparation is completed.

To coordinate the logic executing on two batch units, appropriate global variables (i.e., accessible from any phase) must be reserved for this purpose. This variable may be either of the following:

Discrete value. Often called a "flag," one is required for **Feed Tank A** to signal the **Reactor** that the solution preparation has been completed. Another is required for the **Reactor** to signal to **Feed Tank A** that it is ready to receive the solution.

Numerical value. Usually an integer number, different values can be assigned to indicate the progression of the logic on a batch unit. Using **Feed Tank A** as the example, the states might be as follows:

0 Empty
1 Charging Water
2 Charging Material A
3 Chilling
4 Solution Ready
5 Feeding Reactor
6 Solution Transferred

When the **Reactor** is ready to receive the solution, its logic waits until the state of **Feed Tank A** is **Solution Ready**.

Usually some "handshaking" is required to coordinate the logic. For the block valves in Figure 7.1, assume control is as follows:

1. The block valve on the discharge of **Feed Tank A** is controlled by the logic executing on **Feed Tank A**.
2. The block valve where the solution enters the **Reactor** is controlled by the logic executing on the **Reactor**.

To transfer the solution, both block valves must be open.

The logic executing on **Feed Tank A** opens the block valve on the discharge only while transferring the solution to the **Reactor**. At other times, this block valve must be closed. The logic executing on **Feed Tank A** closes this valve and monitors it for remaining closed.

To initiate the transfer of the solution to the **Reactor**, the following sequence is required:

1. On completion of the solution preparation, **Feed Tank A** indicates **Solution Ready** and waits for the **Reactor** to indicate **Ready for Solution**.
2. When preparation of the **Reactor** to receive the solution has completed, the **Reactor** indicates **Ready for Solution** and waits for **Feed Tank A** to indicate **Feeding Reactor**.
3. When the **Reactor** indicates **Ready for Solution**, the logic on **Feed Tank A** opens the discharge valve, changes its state to **Feeding Reactor**, and waits for the **Reactor** to indicate **Solution Transferred**.
4. When **Feed Tank A** indicates **Feeding Reactor**, the **Reactor** indicates **Transferring Solution** and proceeds to transfer the solution. The **Reactor** can open and close the block valve where the solution enters the **Reactor** in whatever manner is required.
5. When the solution transfer is complete, the **Reactor** indicates **Solution Transferred**.
6. When the **Reactor** indicates **Solution Transferred**, **Feed Tank A** indicates **Solution Transferred** and closes its discharge valve.

In this sequence, steps 1 and 2 can occur in the opposite order, and the remaining logic executes properly. Essentially, **Feed Tank A** and the **Reactor** proceed independently up to step 3 in the above-mentioned sequence. But thereafter, they proceed in a lockstep fashion until the solution transfer is completed.

The above-mentioned logic uses numerical values in globally accessible locations to indicate the status of each batch unit. A comparable result can be achieved using discrete values or flags, but more than one flag is usually required. Potentially one flag could be required for each possible state of a batch unit, but in practice, flags are only allocated for states that are referenced in the logic executing on other batch units.

7.6. SHARED EQUIPMENT: EXCLUSIVE USE

Shared equipment comes in two types:

1. Exclusive use
2. Limited capacity.

This section examines the exclusive use variety. The limited capacity variety is the subject of the next section.

The material transfer systems discussed in a previous chapter are shared equipment of the exclusive use type. Using the single-source, multiple-destination arrangement in Figure 5.3 as the example, this material transport system is restricted to delivering the raw material to only one destination at a given time.

Management of such shared resources requires a mechanism to grant a batch unit the exclusive right to use the raw material transport system for whatever purpose is required. While allocated to one batch unit, no other batch unit is permitted to use the raw material transport system.

To use a shared resource, the logic executing on a batch unit must go through a procedure such as the following:

1. Request the right to use the shared resource. Process types use words such as "Acquire" or "Take" for this action; computer people prefer "Enqueue."
2. Wait until granted permission to use the shared resource.
3. Use the shared resource as desired.
4. Release the shared resource. Computer people prefer words like "Dequeue" over "Release" or "Give."

The logic executing on the batch unit must be well-behaved in two respects:

1. Always acquire the shared resource before using it.
2. Always release the shared resource when finished with it.

There are issues that complicate both of these.

Acquiring the shared resource raises the possibility of a situation known as "deadlock." This can arise when logic executing on two batch units both need the same two shared resources. Consider the following sequence of events:

Batch Unit A
1. Acquires shared resource X
2. At a later time and before releasing shared resource X, requests shared resource Y. However, this shared resource is currently allocated to **Batch Unit B**, so the logic waits for it to become available.

Batch Unit B

1. Acquires shared resource Y
2. At a later time and before releasing shared resource Y, requests shared resource X. However, this shared resource is currently allocated to **Batch Unit A**, so the logic waits for it to become available.

Each batch unit is waiting for a shared resource that is currently allocated to the other batch unit. Neither will proceed until operations personnel intervene.

The potential for deadlock arises when a batch unit requests a shared resource at a time that another shared resource is allocated to that batch unit. One way to prevent deadlock is to impose the restriction that a batch unit can only wait on a shared resource when no other shared resource is allocated to that batch unit. The batch unit can request use of the shared resource, but if its request is not granted immediately (i.e., the shared resource is not free at this time), the batch unit must first release any shared resources allocated to that batch unit and then the batch unit is permitted to wait for the shared resource to be free.

With regard to releasing a shared resource when no longer needed, this is easily incorporated into the logic. However, consider the following sequence of events:

1. The logic executing on a batch unit acquires a shared resource.
2. The logic starts using the shared resource in the customary manner.
3. Before finished with the shared resource, an equipment malfunction causes the logic executing on the batch unit to be suspended.
4. Operations personnel decide to abort the logic executing on the batch unit.

At the time the logic is aborted, the shared resource is allocated to the batch unit. Preferably, the process controls automatically release any shared resources when the logic is terminated, but if not, the user must make provisions to release all shared resources when the logic is aborted.

If the user must provide the logic for allocating a shared resource, care must be exercised. Consider the following approaches:

1. For each shared resource, a globally accessible location is provided to indicate the batch unit to which the shared resource is currently allocated. If not currently allocated to any batch unit, the shared resource is "free."
2. If the shared resource is needed, the logic executing on a batch unit first checks this location to see if the shared resource is free. If not free, the logic repeats the check until the shared resource is free. The resulting loop must not be programmed as follows in a procedural language:

while (shared resource != free)

{

}

This constitutes a tight loop that consumes all available execution time, essentially locking up the process controls until the logic executing on the batch unit is aborted (usually done by the process controls when the logic exceeds its allowable execution time).

while (shared resource != free)

{

    release processor for one scan cycle

}

3. The logic executing on a batch unit is normally activated on a time interval that is designated earlier as the "scan cycle." Such a facility is required to avoid endless loops. Some require specific code to release the processor; some provide special languages that do this "under the hood."

When the shared resource becomes free, the logic writes its identifier into the globally accessible location and proceeds to use the shared resource.

There is another potential problem with the above-mentioned logic. If the resource is free, the following actions should be executed in rapid succession:

1. Check the contents of the global storage location.
2. If the shared resource is free, write the identifier of the batch unit on which the logic is executing into this location.

Real-time software can potentially contain a "bug" that is the result of timing. Starting with the shared resource free, consider the following sequence:

1. The logic executing on **Batch Unit A** retrieves the value of the global storage location.
2. The execution of this logic is suspended, and the logic for **Batch Unit B** begins execution.
3. The logic executing on **Batch Unit B** retrieves the value of the global storage location.
4. The contents of the storage location indicates that the resource is free, so the logic on **Batch Unit B** stores the identifier of **Batch Unit B** in the global storage location and proceeds to use the shared resource.
5. The logic executing on **Batch Unit B** is suspended.
6. The logic executing on **Batch Unit A** is resumed. The contents of the global storage location for the shared resource were previously retrieved. Consequently, this indicates that the shared resource is free, so **Batch**

Unit A stores the identifier of **Batch Unit A** in the global storage location and proceeds to use the shared resource.

With two batch units attempting to use the shared resource at the same time, the results are not predictable but are likely to be bad.

Such bugs are the result of what is sometimes referred to as a "race condition." There is a very narrow window of time where the logic executing on **Batch Unit A** can be interrupted in such a fashion to give the above-mentioned results. Of course, it could also happen for such logic executing on any batch unit. With such a narrow window, the logic is likely to execute properly thousands of time, but given long enough, events such as the previously mentioned will occur. The big problem with real-time software is that there is no way to devise tests that can be guaranteed to detect such problems.

In the literature on computer operating systems, this problem is well known. The solution is to prevent the logic from being interrupted between retrieving the value of the storage location and checking its contents. Most operating systems provide special services whose use will avoid these pitfalls. The key is an uninterruptible "test and set" function that performs as follows:

1. The value of a location is retrieved.
2. If not set, the value of the location is set.
3. A return value indicates whether or not the location was set by the current call.

Before implementing logic for resource allocation, issues such as the previously mentioned must be well understood, which usually means a thorough understanding of the environment in which the logic is executed within the process controls.

A preferable alternative is for the process controls to provide a resource allocation feature. In addition to addressing such issues at the system level, a resource allocation feature can more intelligently allocate the shared resource. The basis for the logic described previously is essentially "if it's free, then take it." If two or more batch units are waiting on a shared resource that is currently allocated to another batch unit, this type of logic makes it difficult to predict which one will acquire the shared resource when it is released. A resource allocation facility maintains a list or queue of the batch units requesting the shared resource and consequently can provide options such as the follows:

1. Allocate in the order requested. That is, first come, first serve.
2. Allocate according to priority. Batch units such as reactors could have a higher priority than the batch units for feed tanks.

One can propose even more elaborate bases for allocating a shared resource, such as increasing the priority of the requesting batch unit during the time that the request remains in the queue. However, the simpler options usually suffice for batch applications.

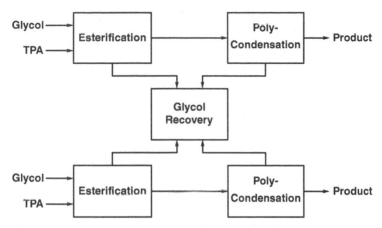

Figure 7.7. Limited capacity shared equipment.

7.7. SHARED EQUIPMENT: LIMITED CAPACITY

Figure 7.7 presents a simplified version of a batch polyester plant. The polyester reaction occurs in two steps:

Esterification. The feeds are glycol and terephthalic acid (TPA). One product of the reaction is water, which is basically boiled off (along with some glycol) and becomes a feed to the glycol recovery unit.

Poly-condensation. The feed is the product from the esterification reactor. One product of this reaction is ethylene glycol, which is also boiled off and becomes a feed to the glycol recovery unit.

The reactions are batch, so the feeds to the ethylene glycol recovery units vary as the reactions progress.

The glycol recovery unit is shared equipment but can be used by all reactors at the same time. However, there is a consideration. Can the flow rate of the combined feeds exceed the capacity of the glycol recovery process? If the answer is yes, the usual approach is to carefully schedule the execution of batches in the reactors.

Delaying the start of a batch is always an option. But once a batch is in progress, delaying is not always feasible. Such a situation arises for the batch digesters in the paper industry. These digesters are basically large pressure vessels for cooking wood. A typical "cook" proceeds as follows:

1. The digester is charged with wood, chemicals, and water.
2. The digester is heated to a specified temperature.
3. The digester is held at this temperature for a specified period of time.
4. Known as "blowing the digester," the pressure is released by venting the vapors within the digester to the vapor recovery unit.

The vapor recovery unit is never sized to permit all digesters to be "blown" at the same time. Again, the "cooks" are scheduled so that the capacity of the vapor recovery unit is not exceeded.

Unfortunately, the best plans occasionally go astray. A problem arises within one or more digesters that cause the cook to be delayed. A potential result is that too many digesters are "blowing" at the same time, which exceeds the capacity of the vapor recovery process. The consequence is that the pressure in the digesters is released more slowly. In effect, this extends the cook. Officially, the cook ends at the start of the blow. However, if wood is sitting in hot chemicals, it is cooking.

If the capacity of the vapor recovery unit is not exceeded, the time for blowing the digester is essentially the same for all cooks, and the cook time can be adjusted accordingly. A sample from each cook is analyzed for a parameter known as the kappa number, which indicates the degree that the wood has been cooked. If the degree of cooking is low, the cook time or cook temperature is increased. If the degree of cooking is too high, the cook time or cook temperature is decreased. In essence, this adjusts for the amount of cooking that occurs while the pressure in the digester is being released.

But when things go astray and the wood in a digester sits in hot chemicals longer than expected, the result is overcooked wood. If a digester is ready for "blowing," but the capacity of the vapor recovery unit is inadequate, the options are rather bleak. Extending the cook time is not an option, so blowing the digester cannot be delayed. However, blowing the digester causes the cook times for all of the digesters currently being blown to effectively increase, the result being that the wood in these digesters will be overcooked to some degree.

The usual approach of addressing such issues is to schedule batches so as to not exceed the capacity of the shared equipment. Provided the batches execute in the usual manner, adhering to the schedule prevents exceeding the capacity of the shared equipment. The problem is that batches occasionally do not execute in the expected manner, sometimes leaving operations personnel with few, if any, options to prevent the capacity of the equipment from being exceeded.

7.8. IDENTICAL BATCH UNITS

Many batch facilities contain multiple items of equipment of the same type. For the batch polyester process illustrated in Figure 7.7, the two esterification reactors are identical and the two poly-condensation reactors are identical. In this context, "identical" means that if a measurement device is installed on one of the units, a measurement device of the same type is installed on the other unit. The same applies to block valves, control valves, control blocks, and so on.

The simple approach to developing the control logic for such units is as follows:

1. Develop the logic for one of the units, using identifiers for measurements, valves, and so on, specific to that unit.
2. Make a copy of this logic and change the identifiers to those for the other unit.

Text editors make this very easy to do. But there is a problem. Any change in the logic must now be implemented in both sets of logic.

For two units or perhaps three, this is manageable, although a little inconvenient. Paper mills that use batch digesters usually have a half dozen or more, all being identical. As the number of identical units increases, modifying the additional sets of logic becomes more than a little inconvenient. Many opportunities are provided to make a mistake—the logic for one of the units either is not modified or is incorrectly modified.

The following approach is preferable, even for only two identical units:

1. Develop the logic for one of the units, using identifiers for measurements, valves, and so on, specific to that unit.
2. Directly activate this logic on another unit, with the process controls effectively substituting identifiers "under the hood."

Having only one set of logic eliminates the need to transfer changes in one set of logic to all other sets of this logic.

This capability is surprisingly easy to implement using a technique called "indirect addressing" that has been available in computer systems since the early days. Instead of directly pointing to something of interest, the code points to a location that contains a pointer to whatever is of interest. This can be readily accomplished using the I/O list from the configuration parameters for a batch unit.

Suppose the desire is to open discharge valve **RCVDISCHA** on batch unit **TANKA**. The code should be written so that it appears that the open applies directly to valve **RCVDISCHA**. But when this code is translated, the result is the index of the location within the I/O list for batch unit **TANKA** that contains a pointer to valve **RCVDISCHA**. This is essentially indirect addressing.

When the logic is executing on batch unit **TANKA**, the I/O list for **TANKA** is used when executing the code. But when the logic is executing on batch unit **TANKB**, the I/O list for **TANKB** is used when executing the code. The same index into the I/O list is used when opening the valve, but the actual valve that opens is **RCVDISCHA** for batch unit **TANKA** but **RCVDISCHB** for batch unit **TANKB**.

In order for this to work, accessing all inputs, valves, and so on, must be through the I/O list for the batch unit. If any device is accessed directly, that specific device will be accessed regardless of the batch unit on which the logic is executing. Occasionally, the same device is used in the logic for all batch units, an example being cooling water supply temperature. However, usually, it is preferable for even these to be accessed through the I/O list for the batch unit.

When configuring the I/O lists for identical batch units, the elements in the lists must be defined properly. For example, if **RCVDISCHA** is element 21 in the I/O list for **TANKA**, then **RCVDISCHB** must be element 21 in the I/O list for **TANKB**.

In addition, a mechanism must be provided to designate which batch units are identical. There may be more than one type of identical units. For the polyester plant in Figure 7.7, the two esterification reactors are identical and the two poly-condensation reactors are identical. However, an esterification reactor is not identical to a poly-condensation reactor, so the logic for an esterification reactor must not be activated on a poly-condensation reactor. One possibility is to designate a type for each batch unit. The logic developed for one batch unit can only be executed on a batch unit whose type is the same as the one for which the logic was developed.

In practice, "identical" batch units often turn out to be "very similar" batch units. Usually, the desire is that they be identical, and when first installed, this may have indeed been the case. But with time, they drift apart for reasons such as the following:

1. A modification is made to some of the batch units but not to all.
2. Something fails and is replaced. The desire may be to replace with exactly the same equipment, but some issues arise. The original equipment may no longer be manufactured, so one can only replace with something that is reasonably compatible. Even if the original equipment can be purchased, the current model may provide desirable features that were not available on the original model.

The dissimilarity between batch units can be addressed in two ways:

1. Declare that the units are not identical and develop separate logic.
2. Incorporate checks into the logic that directs the logic to different paths depending on which batch unit it is executing. Sometimes this can be based on the contents of the I/O list. If a measurement device is installed on some batch units but not on all, a location for this measurement device can be allocated in the I/O list for the batch units. On those batch units without this measurement device, the pointer in the I/O list will be a "null" pointer.

The latter approach increases the complexity of the logic, but unless the batch units are substantially different, is preferred over separate sets of logic.

8

SEQUENCE LOGIC

To accomplish its purpose, most phases must perform a sequence of actions. Hence, the options for implementing a phase must be capable of accepting specifications for the sequence logic and translating these specifications into a form that can be executed by the process controls.

Over the years, various facilities have been provided for implementing sequence logic. Herein the following will be examined:

1. Relay ladder logic (RLL)
2. Standard procedural languages
3. Languages specially designed for implementing sequence logic
4. State machines
5. Grafcet/sequential function charts (SFCs).

The phases required for an application are generally written in one of these.

8.1. FEATURES PROVIDED BY SEQUENCE LOGIC

Executing sequence logic in a process control application requires a number of capabilities, some in common with other applications for sequence logic but some unique to process control.

Control of Batch Processes, First Edition. Cecil L. Smith.
© 2014 John Wiley & Sons, Inc. Published 2014 by John Wiley & Sons, Inc.

8.1.1. Required Features

Regardless of the mechanism for implementing sequence logic, the capabilities include the following:

Arithmetic capabilities. This includes the usual add, subtract, multiply, and divide. But occasionally, functions such as logarithms and exponentials must be computed. Although often provided because they are otherwise available, the trigonometric functions are rarely used in process applications.

Logic capabilities. The entire set of conditional tests are required. Combinations of the tests are frequently used, so combining tests is preferably easy to do. The if-then-else construct or its equivalent is also desirable.

Interface to regulatory control. This must encompass the following:

1. Access to all inputs, both analog and discrete.
2. Operating parameters, such as the mode, set point, and output of a proportional-integral-derivative (PID) control block. For automating batch processes, being able to change set points is not sufficient. In some cases, the PID controller must be switched to manual and the loop output specified directly.
3. Engineering parameters, including PID tuning parameters, alarm limits, alarm arm/disarm, and so on.
4. Discrete control, primarily through the interface to the discrete device driver. The logic must have the capability to specify the desired state of a field device. The logic must also be able to read the transition and invalid status maintained by the discrete device driver for that field device. If one is supported, the sequence logic must have a similar access to the step programmer as described in the previous chapter.

Wait on condition. A convenient mechanism should be provided to suspend execution of the sequence logic until some logic condition is satisfied. Examples include the following:

1. Wait on timer. Both up-counting and down-counting timers are preferably provided, although only one type is sufficient.
2. Wait on the state of a discrete device, such as wait until a rotary switch indicates that a centrifuge is no longer spinning.
3. Wait on the value of a process variable, such as wait until the vessel temperature is less than 10°C.
4. Wait on a response from the process operator. For example, the operator could be asked to confirm that the agitator seal is not leaking.

Access to values from the product recipe. The possibilities include the following:

1. The values from the formula are stored in a structure that is accessible to the sequence logic.

2. Values from the procedure section of the recipe are provided via parameters in the call that activates execution of each phase.

Management of shared resources. Although the allocation of a shared resource can be provided by the logic developed for an application, a more convenient mechanism that addresses all of the issues is very useful.

Coordination between batch units. This can be through either discrete variables (flags) or numerical values (batch unit status) that are globally accessible.

8.1.2. Timed Waits

Except for the wait on timer, all other waits in the previously cited list have one trait in common: it is possible that the wait will never be satisfied, which essentially holds execution of the sequence logic at the wait. This persists until the process operator recognizes that the sequence is not progressing and intervenes.

Consider the wait for the vessel temperature to drop below 10°C. Based on operating experience, an upper limit can be established for the time typically required to satisfy this condition. If the wait has not been satisfied within this time, the process operator's attention is required. Possibilities include the following:

1. Maintenance used hand valves to block off the chilled water, but forgot to open the valves when their work was completed. Most operators can quickly correct this problem, but only once they are aware that a problem exists.
2. The vessel temperature is 11°C but is dropping slowly if at all. The process operator should advise the product technologists of this situation and ask if it is OK to proceed.

With a flexible logic capability, provisions for a timed wait can be incorporated into the application code by incorporating logic such as that in Figure 8.1. In addition to the conditional tests, a timer is also required. The need for a timed wait arises frequently enough that a standard feature for this purpose is useful.

8.2. FAILURE MONITORING AND RESPONSE

The logic for automating a batch process must provide for the following two types of tests, both of which are required in the sequence logic for preparing a solution in the feed tank in Figure 7.2:

Point in time. Once all materials are added, the mixture must be cooled, which is accomplished by phase **Chill** that terminates when the mixer

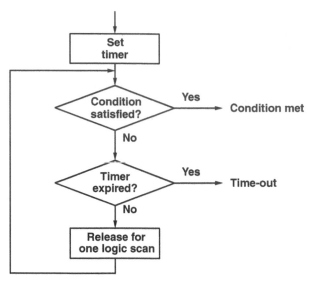

Figure 8.1. Timed wait on condition.

temperature is less then 10°C. Once this condition is satisfied, the sequence logic proceeds, and the test is not repeated.

Repeated test (monitored). Water and material A must be transferred to the mixer, which are accomplished by phases **Add Water** and **Add A**. During these phases, the high-level switch must not actuate. Performing this test only at the start of the phase is inadequate. The test must be repeated throughout the time that the **Add Water** and **Add A** phases (and possibly other phases) are in progress.

8.2.1. Normal Logic and Failure Logic

The objective of repeated tests is to detect when abnormal situations occur and initiate an appropriate response to these situations. In this context, it is useful to divide sequence logic into two categories:

Normal logic. This comprises the sequence logic to execute if all is well; that is, monitoring of level switches and the like do not detect abnormal conditions.

Failure logic. This comprises the sequence logic to execute when monitoring level switches and the like detect an abnormal situation. Depending on the nature of the abnormal condition, different responses may be required.

Sometimes logic is incorporated into the normal logic to respond to certain abnormal conditions. For a reactor, the normal control logic should maintain the reactor temperature within a specified tolerance of its set point. Should the temperature rise above the set point by more than the tolerance, a response to this specific event can be incorporated into the normal logic. In the reactor example, the feeds may be stopped, full cooling may be applied to the jacket, or other appropriate response provided by the normal logic, usually with the expectation that the situation can be resolved and the logic continued.

In a sense, the normal logic is providing the response to a specific abnormal condition. One could propose to handle all abnormal conditions in this fashion. In processes like reactors, the "what if" list becomes very long. In the resulting logic, the normal progression of the reaction is dwarfed by the logic for responding to abnormal conditions.

8.2.2. Defining the Requirements

A "divide and conquer" approach is preferable. First, put on rose-colored glasses and assume "nothing can possibly go wrong." The objective is to defer the myriad of "what if" discussions. Define the logic required to control the process under normal conditions, the resulting logic constituting the normal logic.

The next step is to define the conditions that must exist in order for the normal logic to proceed. This is "what if" from a different perspective. The typical "what if" focuses on a specific event that could potentially occur. But instead of focusing on how to detect that this event has occurred, the objective is to determine the conditions from which it can be concluded that this event has not occurred. If the event has not occurred, then execution of the normal logic can continue.

This approach permits the "what do we do now" discussion to be deferred. The result of this analysis is a list of what must be monitored and for the conditions that must exist so that execution of the normal logic can continue.

8.2.3. Failure Logic

The last step is to define the failure logic. Deferring to the last reflects the reality that the failure logic can easily exceed the normal logic. Before addressing the requirements for failure logic, it is essential to get the normal logic defined.

When an abnormal condition is detected, the following options are available for responding:

1. A response that is specific to that abnormal condition. An example is the temperature excursion discussed earlier for the reactor. Although this could be the result of an equipment issue, the initial response is to assume it is a process issue. Instead of initiating a failure, the logic to address process issues is often incorporated into the normal logic.

2. A response that can be used for a variety of abnormal conditions. Basically, the response to a failure is to place the process in a shutdown state
of some type. The possible shutdown states must be defined, and then
the failure logic must transition the process from its current state to an
appropriate shutdown state. For most reactions, the following (and possibly more) shutdown states are possible:

Hot hold. The feeds are stopped, and the reactor temperature is maintained with the expectation that the problem can be resolved and the
normal logic resumed.

Complete process shutdown. The feeds are stopped and the contents
of the reactor dumped or the reaction appropriately terminated or
"quenched" as quickly as possible. Such actions are usually necessary
on failures such as an agitator malfunction. No recovery is possible—a
vessel with an exothermic reacting medium and no cooling requires
that drastic actions be taken. The product batch has to be aborted and
the problem must be corrected before starting the next batch.

Especially in a complex batch unit such as a reactor, the first approach requires
adding considerable logic to the normal logic, making the result excessively
complex unless restricted to a limited number of abnormal conditions. The
second approach provides a clean separation between the failure logic and
the normal logic. In some cases, a sufficient response to the failure is to drive
the discrete devices to specified states, but in some cases, additional sequence
logic must be provided to appropriately respond to the failure.

The following analogy to a traffic light explains failure detection and
response:

Failure monitoring. Analogous to the traffic light control unit, which determines the color of the traffic light to be either red, yellow, or green.
Green logic. Execute the normal logic.
Yellow logic. Drive the process to a condition such as a "hot hold" from
which recovery to normal logic is possible.
Red logic. Drive the process to a total shutdown condition from which
recovery to normal logic is not possible.

8.2.4. Process Operator Issues

When responding to a failure, a major consideration pertains to the process
operators. Upon initiating the failure logic, the process operator must be
informed. Normally, no operator intervention is required (and may not be
permitted) until the response to the failure is complete; that is, the outputs
have been driven to their failure states and the sequence logic, if any, to
respond to the failure has been executed. In order for the process operator to
correct the problem so that the normal logic can be resumed, the process

operator needs to know what initiated the failure. Specifically, in the list of points monitored to detect a failure, what point in this list caused the failure to be initiated?

Before defining another failure state, one must recognize that adding this state requires that the process operators understand the nature of the new state. As one gives more thought to this requirement, using a currently defined failure state becomes more attractive. Often, these issues arise because using an existing failure state is too extreme. However, one should take the position that the occurrence of the failure must be infrequent. If it occurs too frequently, the options are as follows:

1. Understand why it occurs so frequently and correct this problem.
2. Provide a more appropriate and hopefully less severe response to the failure.

The first option is clearly preferable.

8.2.5. Alarm Issues

When responding to failure conditions, careful attention must be directed to process alarms. Potentially, the actions involved in responding to the failure could cause some process alarms to occur. For example, suppose a low flow alarm is configured for the measurement of the flow through a valve. Closing the valve assures that the low flow alarm will occur.

Such alarms are nuisance alarms and must be avoided. Consequently, as part of the failure response, any alarms that will occur because of the transition to a shutdown state must be disarmed. If no attention is paid to such alarms, the result is likely to be that several alarms will occur each time a failure is initiated.

8.2.6. Step Programmer

The step programmer discussed in the previous chapter can be extended to support failure monitoring and response. The starting point is to separately designate the following within the I/O list:

Controlled outputs. The outputs to these discrete output points are specified for each step of the step programmer.

Monitored discrete points. These must include inputs (such as the high level switch in Figure 7.2), other discrete outputs (such as the other valves on the water header in Figure 7.5), and process alarms.

The definition of the step programmer can be extended to include both failure monitoring and failure states. When only one failure state is defined, the result is three components to the definition of each step of the step programmer:

1. Values for the controlled outputs for normal logic
2. Values for the monitored inputs and outputs to continue with normal logic
3. Values for the controlled outputs for failure logic.

When multiple failure states are defined, items 2 and 3 in the previously mentioned list are required for each failure state.

When the states of the monitored inputs and outputs are different from the desired states, a failure occurs. The step programmer drives the controlled outputs to the values specified for failure, and execution of the normal logic is suspended. Sometimes this is an adequate response to the failure, but in some cases, sequence logic must be activated to appropriately respond to the failure.

8.2.7. Avoiding False Failures

False or nuisance alarms are largely annoying to the process operators. But activating the failure logic unnecessarily leads to lost production time. Short-term transient conditions within the process can cause the monitoring logic to initiate a failure in response to some event whose effect is short-lived. Most processes respond slowly, so a short delay in responding to a true failure often has minimal consequences. Initiating failures on transient events can be minimized by using the following delay times:

Delay time before monitoring. When the step programmer is directed to a step, the monitoring does not commence until the specified time has elapsed. This avoids initiating a failure because of transient conditions associated with events such as starting a pump.

Delay time before declaring a failure. When the monitoring detects a failure condition, the failure condition must persist for the specified time before a failure is initiated. This avoids initiating failures in response to events occurring on other batch units. For example, a pump started by another batch unit can cause transients in piping headers.

8.2.8. Field Device Transitions

Field devices such as valves exhibit travel times. The transition time specified in the configuration parameters for the discrete device driver must allow for this or otherwise the discrete device driver will prematurely declare the device status as **Invalid**. However, there is another aspect of this behavior. Specifying the output to a discrete device is often referred to as "drive." There are two possible interpretations:

Drive-and-wait. Specify the output to the field device and suspend execution of the sequence logic until the discrete device driver confirms that the device is in the specified state. Or in other words, the sequence logic is not allowed to proceed while the device is in transition.

Drive-and-proceed. Specify the output to the field device and continue executing the sequence logic. As a result, some actions are performed while the field device is changing states.

Use of drive-and-wait is most common. In practice, the difference is normally significant only for devices that have a long travel time. One example of such devices encountered in batch plants are the flush-fitting valves that avoid dead spaces at the bottom of a reactor. These valves are motor driven and can have travel times in minutes. While such a valve is opening, events such as venting the reactor can proceed. In such cases, drive and proceed is often appropriate.

Similar issues arise when directing the step programmer to a different step. Drive-and-wait means drive all devices and suspend the execution of the sequence logic until all devices controlled by the step programmer are confirmed to be in their desired states.

8.3. RELAY LADDER DIAGRAMS

Simple sequences have long been implemented by expressing the logic as a relay ladder diagram, and will no doubt continue to be implemented in this manner. The older implementations relied on hardwired logic. Today this has been almost entirely replaced with low-end programmable logic controllers (PLCs). Some of these are very small (referred to as micro-PLCs) with small I/O counts, such as 16 discrete inputs max and 4 discrete outputs max.

8.3.1. Gas-Fired Furnaces

An example of a simple sequence is the ignition sequence for a gas-fired residential furnace. Ladder logic can easily implement sequences such as the following:

1. When the start button is pressed, start the blower and ignite the pilot.
2. Open the main gas valve when the following are satisfied:
   (a) The blower has run for the specified time (to purge the furnace).
   (b) The thermocouple confirms that the pilot flame is present.

The relay ladder diagram to accomplish this is illustrated in Figure 8.2.
This relay ladder diagram consists of two rungs or networks:

Rung 1: Input **X1** is from the thermostat and is the on/off control input for the furnace. When **X1** is energized, the pilot valve is opened, the blower is started, and a delay-on output relay is activated.

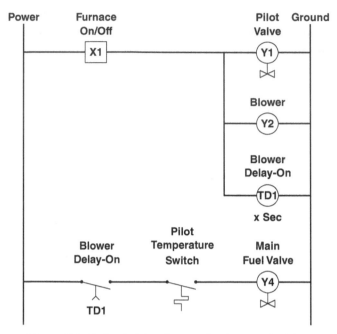

Figure 8.2. Relay ladder diagram for furnace ignition.

Rung 2: The main fuel valve opens when the following are true:
1. The delay-on output relay is energized, indicating that the blower has been running for the specified time.
2. The temperature switch on the pilot closes, indicating that the pilot is lit.

The engineering organizations in most process companies provide a drawing illustrating the symbols to be used in their relay ladder diagrams, but most are more similar than different.

A relay ladder diagram is actually a representation of discrete logic. However, simple, repetitive sequences can be readily expressed using a relay ladder diagram and then implemented either with hardwired logic or within a PLC. The logic in Figure 8.2 only represents the basic sequence. A sequence is required on furnace shutdown, and the logic must also incorporate a contact on the door, a temperature switch to detect flame roll-out, a temperature switch to detect furnace overheating (the blower may not actually be running or the air is blocked), and so on.

8.3.2. Issues with Relay Ladder Diagrams

As a sequence becomes more complex, the complexity of the relay ladder diagram increases rapidly, making discrete logic implementations of sequence logic more challenging. In addition, the following issues arise:

1. PLCs and their hardwired logic predecessors are within the domain of the electrical department, which would normally implement repetitive sequences such as the one in Figure 8.2 for the furnace. The sequence in Figure 8.2 requires no analog capabilities. The PLCs installed by the electrical departments are mostly low-end PLCs with largely, if not exclusively, discrete I/O.

2. The sequences required for batch automation are within the domain of the instrument and controls group. Although most applications are dominated by discrete I/O and discrete logic, batch automation almost always involves some analog I/O (certainly key temperatures) and analog logic (such as PID control). The typical instruments and controls group favors distributed control system (DCS) products that can be purchased from the traditional suppliers of such measurement and control equipment. The instruments and controls group is not fond of relay ladder diagrams, and should they choose to go in that direction, a turf battle with the electrical department could ensue.

The previously cited debate actually involves apples and oranges. The low-end PLCs are not capable of batch automation, and these products are not viable competitors to DCSs. To address the requirements of this market segment, the PLC manufacturers enhanced their products by adding analog I/O, PID control, and numerous other features found in DCSs. The result is an upper-end PLC that is a viable competitor to a DCS. Although most retain the capability of being programmed in ladder logic, most provide additional facilities for implementing sequence logic.

8.3.3. Issues with Electricians

Ladder diagrams such as the one in Figure 8.2 are usually created by the engineering department. Electricians are expected to understand such diagrams and troubleshoot problems arising in the field, the most common problem being a faulty input. As normally drawn, power is supplied at the left of each rung, and power must flow through a circuit (or rung) to ground to energize an output. In hardwired circuits, electricians are taught to check the "power flow." Starting from the left, the electrician tests for the presence of power at each connection point within the rung until a connection with one of the following is located:

1. Power is present when it should not be.
2. Power is not present when it should be.

Graphical programming tools make this even easier by displaying each rung with an indication of where power is present. Most electricians quickly mastered the graphical tools for troubleshooting rungs of ladder logic.

Because of this capability on the part of the electricians, companies often require that simple sequences such as the one in Figure 8.2 be represented as ladder diagrams and implemented accordingly. Alternatives to ladder logic have been available for some time. However, electricians have traditionally been expected to work with ladder diagrams only. Especially in the United States, the inability of the electricians to troubleshoot anything except ladder diagrams has proven to be a major deterrent to the acceptance of any alternative to ladder logic.

8.4. PROCEDURAL LANGUAGES

From the very beginning, digital control systems permitted programs to be written in languages such as FORTRAN, Basic, C, C++, and so on, to retrieve the values of inputs (in engineering units), to specify control valve openings, to open/close block valves, and so on. The capabilities were generally provided through subroutines or functions that could be called from the procedural language.

8.4.1. Program Structure

To implement sequence logic, the program is usually structured as illustrated in Figure 8.3. The program is divided into short segments that will be referred to as "steps." Again, the term "step" is used in various ways, and "step" as used for a program is not necessarily the same as a "step" as used in the previous chapter for the martini mixer.

The program is activated on a fixed time interval, such as 1 second. Each time the program is activated, it executes the portion of code for one step. The steps are assigned an identifier, usually a numerical value. The number of the

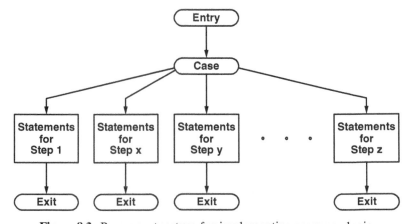

Figure 8.3. Program structure for implementing sequence logic.

step to be executed is preserved between executions. When first activated, the program executes a predetermined step, usually step 0 (in C or C++) or step 1 (in Basic or FORTRAN). Each language provides a case statement (or its equivalent) that directs execution to the appropriate step. When conditions are satisfied for proceeding to another step, the currently executing step changes the value of the step before exiting.

8.4.2. Example

To illustrate, suppose the program is currently executing step 21, which is maintained as variable **Step** within the program. The temperature of a mixer is available as variable **TtMixer** within the program. When **TtMixer** is below 10°C, the sequence is to proceed to step 31. The code that immediately follows will treat 10°C as a constant, but in practice, the value for the temperature target should originate in the product recipe.

In a procedural language such as FORTRAN, such logic would normally be expressed as follows:

```
21 IF (TtMixer .GT. 10.0) GOTO 21
   GOTO 31
```

This approach is not acceptable. If **TtMixer** is greater than 10°C, the aforementioned **IF** statement is an endless loop that consumes all available processor execution time.

Using the structure illustrated in Figure 8.3, the endless loop is avoided with the following coding:

```
21 IF (TtMixer LE. 10.0) THEN
      Step = 31
   ENDIF
   STOP
```

If **TtMixer** is greater than 10°C, step 21 terminates without changing the value of **Step**, so execution of step 21 will be repeated on the next activation of the logic. If **TtMixer** is 10°C or less, the value of **Step** is changed to 31, so step 31 will be executed on the next activation of the logic.

8.4.3. Issues with Procedural Languages

The previous examples in FORTRAN are admittedly somewhat cryptic. However, judicious use of features such as **PARAMETER** statements will greatly enhance the readability of the code. In this respect, C and C++ are even better, providing features such as the **enum** statement. With modest effort, the code can be written so as to be very readable.

Efficiently supporting a procedural language requires a good development environment, a sophisticated operating system, and software personnel who know how to use them. This raises the following issues:

1. Operating companies fear software development. This is a legacy of the bad experiences from the early days of digital controls. The early projects experienced long delays and large cost overruns (took twice as long and cost twice as much). But despite this, most projects eventually made a lot of money.
2. With regard to computing technology, software people find the process industries stogy and "long in the tooth." Smart phones, social networking, and the like are far more dynamic and exciting.

8.5. SPECIAL LANGUAGES

At one time this was the preferred approach offered by most DCS suppliers, although many now offer SFCs that will be discussed in a subsequent section. The initial offerings were derived from Basic; the later offerings were derived from C. These special languages were not as feature-rich as the procedural languages from which they were derived, and in practice, some of their features are not required to implement sequence logic.

8.5.1. Parsing Procedural Languages

Especially when initially developed, one objective was to provide statements that could more easily be parsed (i.e., translated). To understand the difficulty of parsing a feature-rich language such as FORTRAN, consider the following statement:

```
DO  50  J  =  1.5
```

Spaces are immaterial in FORTRAN statements, so this statement is the same as

```
DO50J=1.5
```

This statement is most likely a **DO** statement with a typo, that is, "1.5" should really be "1,5." But a good FORTRAN compiler will be perfectly happy with it. The compiler understands this statement to mean "assign the value 1.5 to the variable **DO50J**." Some compilers will notice that variable **DO50J** does not appear elsewhere in the program and at least issue a warning. Some FORTRAN programmers stipulate strict typing (i.e., require explicit declarations for all variables), and if so, the compiler will generate an error indicating that **DO50J** has not been declared.

Basic is simpler to translate because the nature of the statement can be determined from the first few characters in the statement. Early versions of Basic required arithmetic assignment statements to begin with "LET," an example being the following:

```
LET DO50J = 1.5
```

Later versions somewhat relaxed this. If the nature of the statement cannot be determined from the first few characters, it is assumed to be an arithmetic assignment statement. As a result, parsing Basic statements is much easier than parsing FORTRAN statements.

8.5.2. Access to Real-Time Data

Instead of using subroutines or functions to obtain values of inputs, to specify control valve openings, to open/close block valves, and so on, most special languages accomplish these through enhancements to the language. Some permitted the tag names of process points to explicitly appear within the statements. Some provided specific statements for opening valves, starting pumps, and so on.

8.5.3. Avoiding Endless Loops

Some special languages also attempted to prevent users from getting into trouble. In a sense, the standard procedural languages give the user lots of flexibility or "lots of rope." Unfortunately, occasionally, someone will end up with this rope around their necks by coding statements such as the following:

```
21 IF (TtMixer .GT. 10.0) GOTO 21
```

A special language can avoid the endless loop by executing the **GOTO** in the following manner:

1. **GOTO**s that jump to a subsequent statement within the program are performed immediately.
2. **GOTO**s that jump to the current or a previous statement cause the program to be suspended on the current execution cycle. The target statement of such **GOTO**s is executed when the program is resumed on the next execution cycle.

While such interpretations offer advantages, the programmers need to appreciate what goes on "under the hood."

8.5.4. Long-Term Issues

Languages such as Basic, C, C++, FORTRAN, and so on, are widely known, will be supported in the future, and will evolve by incorporating advancements in computing technology. Despite the advantages offered by a special language, their future survival and evolution are not assured. By computer standards, the process control market is miniscule. Supporting any custom product developed for this market is problematic. Without continuous efforts toward product improvement, any product becomes "long in the tooth," especially in a rapidly evolving industry such as computing technology.

8.6. STATE MACHINE

The drum timer described in the previous chapter is a crude implementation of a state machine, or more precisely, a finite state machine. Each step on the drum timer sets the outputs to specified values, which should drive the process to a certain state. For the sequence logic for the martini mixer described in the previous chapter, Figure 8.4 represents the logic in the form of a state diagram. A state machine executes such logic by progressing from one state to another.

For process applications, there are two essential components for executing a state diagram:

State. For each state, the outputs to the process are driven to specified values. This aspect of the drum timer is exactly what is required.

Transition. Logic is required to determine when to proceed from the current state to another state. State machines can direct execution from the current state to any state, including the current state. The logic may direct execution to alternate states depending on conditions within the process.

With respect to the state diagram in Figure 8.4 for the martini mixer, state machines are far more flexible than drum timers with regard to the following aspects of transitions:

1. Proceeding to another state can be initiated for reasons other than elapsed time.
2. States do not have to be executed sequentially.

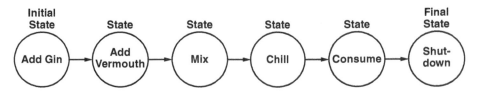

Figure 8.4. State diagram for martini mixer.

8.6.1. Special States

A state machine treats two states in a special way:

Initial state. When activated, the state machine starts with the initial state. For the martini state diagram in Figure 8.4, the initial state is **Add Gin**.

Final state. When directed to the final state, the state machine terminates execution of the sequence logic. There is no logic for exiting the final state. For the martini state diagram in Figure 8.4, the final state is **Shutdown**.

There can be only one initial state. At least conceptually, there could be more than one final state. But imposing the restriction of only one final state means that the process will always be in a predetermined state when the state machine terminates execution. In most applications, this behavior is desirable.

8.6.2. Transitions

State machines support alternate paths, such as illustrated in Figure 8.5. From State 2, the transition can be to either State 3a or State 3b. The logic for State 2 determines which one should be the next state. Regardless of the path taken, execution eventually arrives at State 6, from which a single path leads to the final state.

Loops are permitted in a state diagram. Figure 8.6 presents an alternate state diagram for a martini mixer. The transition from **Shutdown** proceeds to **Add Gin**. A martini mixture is prepared on each execution of the loop. As there is no final state, the state machine executes until terminated manually.

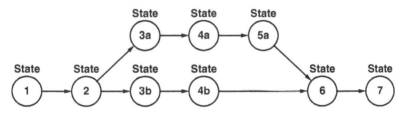

Figure 8.5. State diagram with alternate paths.

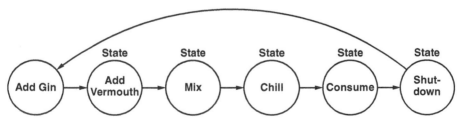

Figure 8.6. Loop in state diagram for martini mixer.

However, it would be relatively easy to provide two transitions out of the **Shutdown** state, one to **Add Gin** to make another martini mixture and another to a final state. In one sense, this is preferable—terminating the logic from any state other than **Shutdown** is unwise.

State machines do not provide for parallel paths. In the state diagram in Figure 8.6b, the two paths are alternate paths. Only one will be executed. There are no constructs within the state diagram to specify that both must be executed. However, two state machines can be executed simultaneously. To provide parallel execution, one state machine must activate another state machine.

8.6.3. Excessive Number of States

In some process applications, a surprisingly large number of states arise. Figure 8.7 illustrates a co-feed involving four streams. Each feed is terminated when the specified amount of material has been transferred. Under ideal conditions, all four feeds would terminate at exactly the same instant of time. But in practice, some variation occurs in the time that each feed is terminated. Furthermore, the order in which the feeds terminate is not known, with every combination being a possibility.

If the feeds are stopped by proceeding from one state to another, the state diagram in Figure 8.8 is required. The sequence of states begins with all feeds in progress; the sequence terminates with all feeds stopped. Between these two states are 14 possible intermediate states.

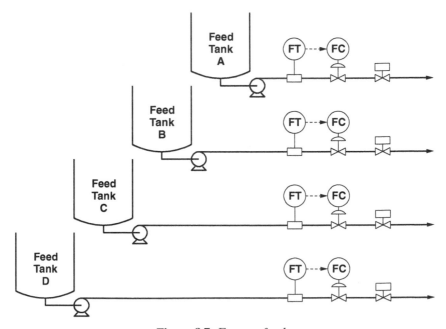

Figure 8.7. Four co-feeds.

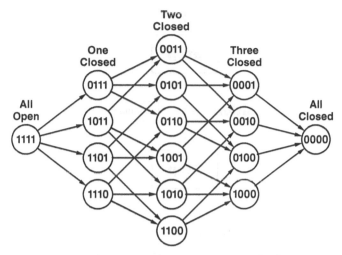

Figure 8.8. State diagram for four co-feeds.

Although a departure from a true state machine, one possibility is to provide a mechanism to stop a feed without progressing from one state to another. When the first feed must be stopped, the state diagram proceeds to an intermediate state for which all feeds are declared as **Don't Care**. The logic for this state stops the feeds individually until all are stopped, and then proceeds to the state with all feeds stopped.

8.7. GRAFCET/SEQUENTIAL FUNCTION CHARTS (SFCs)

Evolving from a prior methodology known as Petri nets, Grafcet is a graphical method for specifying and documenting the logic in controllers. Being developed in France in 1977, a Google search on "Grafcet" will yield many articles written in French. Grafcet quickly evolved to a very effective graphical programming tool generally referred to as SFCs. Two IEC standards pertain to Grafcet:

IEC 60848 (2013) "Specification language GRAFCET for sequential function charts"

IEC 61131-3 (2013) "Programmable controllers—Part 3: Programming Languages"

The latter is not restricted to SFCs; other alternatives (including ladder diagrams) are included.

When applied to controllers, a major limitation of traditional flowcharts and state diagrams is the inability to express two sequences that must be performed at the same time, that is, in parallel. This capability existed in Grafcet from the very beginning.

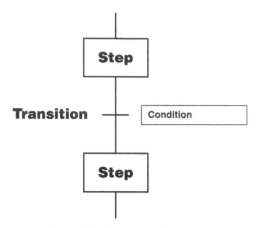

Figure 8.9. Step-transition-step.

8.7.1. Step-Transition-Step

The basic construct in the graphical representation is referred to as step-transition-step and is illustrated in Figure 8.9. The nature of a step and a transition are as follows:

Step. A step consists of actions to be performed. These can be expressed by statements similar to those in procedural languages. Actions may be open/close valves, change set points of controllers, and so on.

Transition. A transition is a condition. It may be a simple condition, but logical operators (AND, OR, XOR) may be freely used to compose a complex condition. The condition for the transition may simply be TRUE, meaning that the transition is immediately executed.

The progression of the logic is as follows:

1. Execute the actions in one step.
2. Wait until the condition for the transition is satisfied.
3. Execute the actions in the next step.

Other constructs are available to provide for alternate paths and parallel paths.

8.7.2. Initial Step and Terminal Step

The special representations used for the initial step and the terminal step are presented in Figure 8.10. The representation for the terminal step clearly suggests an electrical orientation for the developers of Grafcet/SFCs.

The path out of the initial step is immediately followed by a transition. There is no path into the initial step.

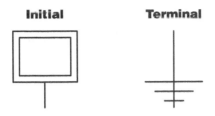

Figure 8.10. Initial step and terminal step.

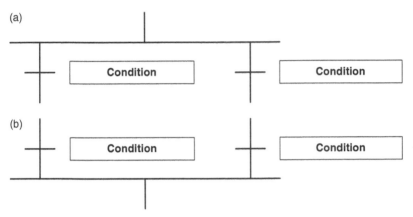

Figure 8.11. Divergent OR/Convergent OR: (a) Divergent OR. (b) Convergent OR.

The terminal step is immediately preceded by a transition. There is no path out of the terminal step.

8.7.3. Divergent OR/Convergent OR

The Divergent OR is used when only one of two (or more) paths is to be executed. As illustrated in Figure 8.11a, there is only one path into the Divergent OR, but there may be two or more exit paths, each beginning with a transition. The condition specified for the transition determines which path will be executed. If none of the conditions are TRUE, the logic waits at the Divergent OR, and the conditions are checked repeatedly. When one of the conditions becomes TRUE, that path is executed.

The Convergent OR is used when the logic may arrive at a step from one of several possible steps. As illustrated in Figure 8.11b, there is only one exit path from a Convergent OR, but there may be multiple entering paths. Each entering path terminates with a condition. When the condition on any one of the entering paths is satisfied, the logic proceeds to the first step of the exit path.

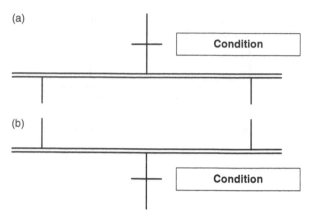

Figure 8.12. Divergent AND/Convergent AND: (a) Divergent AND. (b) Convergent AND.

8.7.4. Divergent AND/Convergent AND

The Divergent AND is used when two (or more) paths are to be executed simultaneously (or "in parallel"). As illustrated in Figure 8.12a, there is only one path into the Divergent AND, but there are two or more exit paths. The Divergent AND is immediately preceded by a transition. When the condition specified for this transition is satisfied, execution commences for all exiting paths from the Divergent AND.

The Convergent AND is used when two (or more) paths being executed in parallel must complete before the logic proceeds to another step. There are multiple input paths to the Convergent AND, but only one exit path that begins with a transition. The logic proceeds to the next step in the exit path provided the following are true:

1. All input paths have been executed up to the Convergent AND.
2. The condition for the transition for the exit path is satisfied.

8.7.5. Product Receiver

To illustrate the use of the Convergent OR, the control logic will be developed for the product receiver in Figure 8.13. Product is retained in the product receiver until it is "full," with the interpretation of "full" being that the high-level switch is actuated. The storage valve is opened to transfer the product to storage. The storage valve remains open until the product receiver is "empty," with the interpretation of "empty" being that the low-level switch is not actuated.

Beginning with the storage valve closed, the logic proceeds as follows:

1. When the high-level switch **HL** indicates **High**, open the storage valve **RCV**.

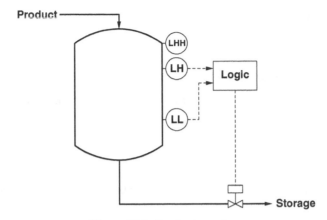

Figure 8.13. Product receiver.

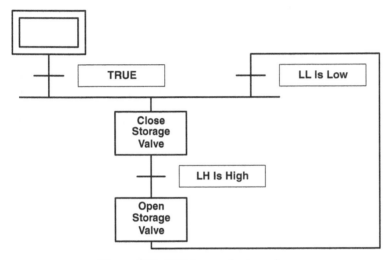

Figure 8.14. SFC for product receiver.

2. When the low-level switch **LL** indicates **Low**, close the storage valve **RCV**.

This logic is embodied in the SFC in Figure 8.14. A Convergent OR is used to create a loop that consists of the following steps and transitions:

Step: Close storage valve **RCV**.
Transition: Wait until high-level switch **HL** indicates **High**.
Step: Open storage valve **RCV**.
Transition: Wait until low-level switch **LL** indicates **Low**.

When the logic is activated, the logic enters the loop at the step that closed the storage valve.

8.7.6. Discrete Logic Implementation

An alternative is to implement using discrete logic. Only one statement is required:

`RCV = (RCV and (LL is OK)) or (HL is High)`

Storage valve RCV will be open under the following two conditions:

1. Storage valve **RCV** is currently open, and the low-level switch **LL** indicates that the level is **OK** (or conversely, does not indicate **Low**).
2. The high-level switch **HL** indicates **High**.

This single logic statement can be implemented in a single rung or network of RLL.

Some will correctly observe that the discrete logic implementation is more concise and can be executed more efficiently than the SFC in Figure 8.14. However, most process types find the SFC representation in Figure 8.14 far more "readable." Those with extensive experience with relay ladder diagrams would not agree, but the reality is that few process types are into relay ladder diagrams to that degree. Furthermore, given the capabilities of today's digital control systems, computational efficiency is no longer paramount.

8.7.7. Co-Feeds

Figure 8.7 illustrates the process with four co-feeds used in the previous discussion pertaining to state machines. Figure 8.15 illustrates the SFC logic for the co-feeds. When the time arrives to start the co-feeds, a Divergent AND activates four parallel paths, one for each co-feed. Each path consists of the following steps and transitions:

Step. Start the co-feed.
Transition. Wait until the total for the co-feed is attained.
Step. Stop the co-feed.

When all four paths have completed, the Convergent AND proceeds with a single path of logic.

As discussed previously, discrete devices can be operated via the following two approaches:

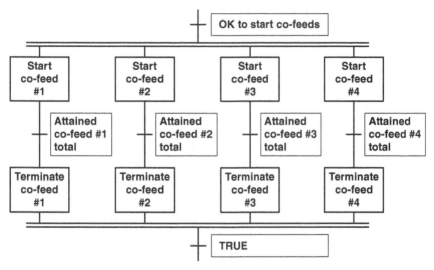

Figure 8.15. SFC for co-feeds.

1. Operate each device individually.
2. Operate the devices via the step programmer described in the previous chapter.

The logic in Figure 8.15 assumes that the devices can be operated individually. If the devices can only be operated through a step programmer (or its equivalent), the number of steps required for the step programmer is the same as for the state machine presented in Figure 8.8.

9

BATCHES AND RECIPES

The objective of a product batch is to manufacture a certain amount of a material as directed in the product recipe for that material. The "product" may be a salable product, or it may be an intermediate that becomes the feedstock to another production process.

The main requirement for a recipe is to provide all of the product-specific information required to manufacture a batch of that product. However, consideration must be given to associated activities such as the following:

Scheduling. Most plants designate an individual to be responsible for determining what product batches are to be manufactured. The industry is evolving from entirely manual scheduling procedures to schedules being generated by the MRP (Material Requirements Planning or Material Resource Planning) systems at the corporate level.

Production control. Most production facilities are not dedicated to producing one product batch at a time. Instead, several product batches are simultaneously in some stage of production. The more complex the production facility, the greater the opportunity for mistakes.

Historical records. Focusing completely on "getting the pounds (kilos) out" is "old school." Capturing the operating data during production of the batch has become essential, and in some cases, is a requirement mandated by regulatory agencies. The original incentive was to better respond to customer complaints, which made data collection and storage

Control of Batch Processes, First Edition. Cecil L. Smith.
© 2014 John Wiley & Sons, Inc. Published 2014 by John Wiley & Sons, Inc.

a pure cost. But returns can be generated by analyzing these data to improve scheduling, to better utilize equipment, to alert coming problems, and so on.

9.1. ORGANIZATION OF RECIPES

Manufacturing a product batch entails activating an appropriate sequence of phases on one or more batch units. Specifically, the following must be driven by the product recipe:

The specific phases to be activated and the order in which they are activated. For manufacturing a particular product, the possibilities include the following:
1. Some phases may be activated more than once.
2. Some of the available phases are not required when manufacturing this particular product.

The values of the parameters required to execute each phase. The parameters may be either numerical values, discrete values (flags), or character strings.

Product recipes in a flexible batch facility must be more than a set of data. Somehow the recipe must direct the execution of the phases on each batch unit used in the manufacture of the product batch.

9.1.1. Origin of Product Recipes

Control specialists do not write recipes. In the chemical industry, recipes are written by product chemists. In the food industry, recipes are written by food technologists. In every case, the recipes are written by someone intimately familiar with the technology. Rarely do these individuals have much background in control. The role of control specialists is to translate the product recipes from their original form to a form acceptable to the process controls.

For most major product lines, a development group is tasked with the responsibility of enhancing the product line. Most have smaller versions of the production equipment that permits them to experiment with new products and to produce a sufficient quantity that potential customers can evaluate the product. In some cases, this development group is physically located at a plant site, and will sometimes use the production equipment should a customer require a substantial amount of the product for evaluation.

Such groups are the origin of most new recipes. Most have a traditional approach for writing recipes, which is often very different from the form required by the process controls. Product chemists usually write recipes in "long-hand," such as illustrated by the example in Table 9.1.

TABLE 9.1. Example of a Recipe

Raw Materials
    1820 kg of A
    740 kg of B
    20 kg of C
    320 kg of water

Solution Preparation Instructions (Feed Tank)
    1. Add water.
    2. Add solute C.
    3. Mix until solute is dissolved.
    4. Cool to 10°C.

Reaction Instructions (Reactor)
    1. Add B.
    2. Start agitator.
    3. Adjust temperature to 30°C ± 2°C.
    4. Feed A and the solution from the feed tank must be fed simultaneously at the ratio of 0.187 kg solution per kg A.
    5. Initiate the reaction by starting the co-feed at 5 kg/min.
    6. Wait for the exotherm. If no exotherm within 10 minutes, stop the feeds.
    7. Ramp the feed rate for A to 12 kg/min over 10 minutes. Allow temperature to rise to 65°C and then maintain at this temperature until feeds complete.
    8. Continue to feed A at a rate of 12 kg/min until all A has been transferred to the reactor. Make sure that all of the solution from the feed tank has been transferred to the reactor.
    9. And so on.

9.1.2. Organization of Product Recipes

Typically, the raw material quantities are separately listed, along with anything else that is required to manufacture the batch. The idea is one can verify that all of the required materials are available before attempting to manufacture the batch. This is analogous to the function provided by the corporate-level MRP system for production operations.

The recipe in Table 9.1 contains two activities that can be performed simultaneously (or in parallel), one for preparing the solution in the feed tank and the other for carrying out the reaction. As written in the recipe from the development group, activities that can be performed simultaneously are separately listed. This conforms nicely with the concept of operations in the context of structured logic for batch automation.

Within each operation, the numbered steps usually correspond to individual phases. The details of how to perform each step is very dependent on the equipment to be used, but none of this is reflected in the recipe in Table 9.1. These details are incorporated into the phase that performs that step.

9.1.3. Phases in an Operation

Execution of each phase requires parameters, such as the amount of raw material to transfer, the initial feed rate for the reaction, and so on. In the recipe in Table 9.1, values for these parameters may be from two sources:

From the formula or raw material list. The amounts of each raw material are specified in the raw material list in the recipe in Table 9.1.

From the statements within the instructions. The initial feed rate for the co-feeds is specified in the statement for one of the numbered steps.

For the recipe for the process controls, one possibility is to extend the raw material list to contain the values of all parameters. With this change the term "formula" is more appropriate than "raw material list." However, there are downsides to this approach:

1. It imposes more work on the person who translates the recipe from the process chemist to the recipe for the process controls.
2. It increases the opportunities for mistakes.
3. It lessens the "readability" of the recipe.

Activating the phases that comprise an operation in the product recipe is akin to a sequence of subroutine calls in a procedural program. For example, the solution preparation operation for the recipe in Table 9.1 could be translated to the following sequence of subroutine calls:

```
CALL  AddWater(320.0)      (Add water)
CALL  AddSolute('C',20.0)  (Add solute C)
CALL  Mix(5.0)             (Mix until solute is dissolved)
CALL  CHILL(10.0)          (Cool to 10°C)
```

However, writing a procedural program that is specific to an operation within a specific product recipe is not a good approach. Instead, the call sequence must be generated dynamically from the information in the recipe for the control system. However, one could envision statements such as the following:

```
AddWater P1=320.0
AddSolute P1='C' P2=20.0
Mix P1=5.0
Chill P1=10.0
```

The first statement is understood to mean activate phase **AddWater** and provide the value 320.0 for the first parameter for the phase.

Three approaches are possible when translating statements such as the above:

Interpreter. Each statement is parsed, and the specified actions are performed immediately. Execution is inefficient because the statement must be parsed each time it is executed, which usually imposes a simple syntax on the statement. However, run-time errors explicitly identify the statement (it is always the one most recently parsed).

Pre-processor. The statements are translated to a pseudo-code or macro-code that is subsequently executed by an interpreter. One advantage of this approach is that all syntax errors are issued before execution commences (a true interpreter would not detect syntax errors until the first attempt at executing the statement). Run-time errors are usually very explicit, especially if a copy of the original statement is retained along with the pseudo-code.

Compiler. The statements are translated to code that can be directly executed by a computer. Execution is most efficient, and languages such as FORTRAN and C++ are usually translated using a compiler. A common criticism is that run-time error messages are cryptic, but this not always true. Modern compilers provide enough information that the original statement can be easily ascertained.

With modern computers, execution efficiency is not a major concern, making all of the above approaches viable. The decision as to which of the above approaches is used is largely at the discretion of the supplier of the process controls. Issues of importance to users include the following:

1. Available features
2. Resolution of errors, especially during run-time
3. Portability, both to a subsequent version of the processes controls from the current supplier or perhaps to an entirely different supplier.

9.2. CORPORATE RECIPES

Initially, the primary focus for the product recipe was on manufacturing a product batch. But today, a distinction is generally made between the following two types of recipes:

Corporate recipe. By using a "standard size" (such as 1000 kg of product), this recipe is independent of the size of the equipment that will actually be used to manufacture each batch of product.

Site recipe. The quantities in this recipe are specific to the equipment that will be used to manufacture each batch of product.

The historical evolution cannot be ignored. Most plants were using site-specific recipes long before the concept of a corporate recipe arose. Plants currently using site recipes are not especially excited about switching to corporate recipes. Switching to the corporate recipe consumes scarce resources, but in addition, the "if it ain't broke, don't fix it" mentality kicks in.

9.2.1. Tailoring Recipes to a Site

Anyone with experience is well aware of the pitfalls of maintaining duplicate copies of what should be the same information—discrepancies arise, even with the best of intentions. Furthermore, plants want the option to "tailor" the recipe, also usually with the best of intentions. Potential motivations include the following:

1. Plant management is always receptive to anything that improves plant productivity. As an example, suppose the corporate recipe adds A and then B. At this plant, this sequence entails a flush after both additions. But if B is added followed by A, no flush is required between B and A. Eliminating a flush shortens the time to manufacture a batch, which increases plant productivity. Do such changes affect product quality? The process technologists usually have an answer to this question, but with no prior experience with the reversed order, their answers are ultimately opinions.

2. Process operators are always receptive to anything that makes their work easier. Given the large number of raw materials utilized in many specialty batch facilities, manual additions for certain materials are unavoidable. Especially for vessels located at some distance from the control room, operators prefer to make one trip rather than two. If two manual additions are required, the temptation to do both in one trip is hard to resist, even if this means changing the order of adding materials.

This is no different from cookbook recipes. My mother always knew a better way, so she routinely "improved" the recipe in the cookbook. And she could never understand why the final product was not quite right, but that it might be the result of her enhancements never seemed to occur to her!

9.2.2. Versions of a Product Recipe

The counterpart to such practices in specialty batch facilities can lead to three versions of a product recipe:

Corporate recipe. Being the "official" method for manufacturing a product, one possibility is to declare the product recipe as produced by the product development group to be the corporate recipe.

> **Site recipe.** Tailoring the corporate recipe to a site gives the method to follow at that site when manufacturing the product.
>
> **Actual recipe.** Although rarely given a name, the "actual" recipe is the method the process operators use to manufacture the product.

One objective of automating a batch process is to eliminate the latter category, but "greatly reduced" is often the practical result. Some people just always know a better way, again with the best of intentions.

In a manufacturing process, such practices are big headaches for those at the corporate level with responsibility for the product line. With customers, they would like to take the position that we have a standard way that we make this product, and all plants do it that way. A product that is made in plants at different locations is packaged in the same containers and labeled the same way. The hope is to avoid the following situations:

1. A customer reports that sometimes the product performs as expected but sometimes not.
2. On further investigation, it is discovered that only the product from plant A is performing as expected.

Will orders for the product be accepted when accompanied by the stipulation that the product must be manufactured at plant A? When the plants are not operating at full production, refusing orders is unlikely.

Are such situations more likely today than in past years? There are two opposing factors:

1. The quality specifications for the manufacturer's products have narrowed, meaning that the product is more consistent. This makes situations such as the above less likely.
2. The same is occurring for the customer's production process. The era of forgiving chemistry is over. Consequently, even small variations in the quality of materials consumed in the customer's production process could have consequences that are detected.

9.2.3. Issues with the Corporate Recipe

While there are definitely some appealing aspects, using the corporate recipe at a plant site raises more issues than initially suspected. Examples of issues that can arise include the following:

1. A raw material may be available in different forms, different concentrations, and so on. The usual way to address this one is to state the amount of a pure or "standard" material, from which the plant site computes actual quantities. For an acid, the corporate recipe would state the amount of acid of a specified concentration. If the acid is available in a

different concentration at the plant site, the amount of the acid feed is adjusted, with the difference offset by adjusting the amount of a water charge. Seems simple, but issues can arise. If the water charge is to provide the heel in the reactor, a lesser amount of water might not permit the agitator to be started.

2. An alternate raw material can be substituted for the one specified in the recipe. The analogy in cookbook recipes is substituting oleo for butter. When the alternate raw material is available locally and at a lower cost, the plant may use it for every batch. When prices and availability fluctuate, the plant will use the alternate for some batches but not all. For this one, there are ramifications for the MRP system.

3. In a pulp digester, a trade-off can be made between time and temperature using a quantitative relationship for a parameter called the H-factor. To increase throughput, increase the temperature of the cook, which permits a shorter cook time to achieve the same degree of cooking.

4. For an atmospheric reactor outfitted with a reflux condenser, the temperature depends on the altitude where the plant is located. For example, the pressure and temperature in the reactor in the Denver facility are lower than in the Houston facility. All reactions are affected by temperature, but reaction rates are also affected by the concentrations of reactants and products. This certainly affects reaction times, but can also affect product yields. But unlike the H-factor for a pulp digester, rarely are quantitative relationships available. Traditionally, product specialists compensate for such effects by making small adjustments in reacting conditions, usually on an individual product basis.

5. Equipment size has ramifications beyond just scaling the quantities in the standard recipe. Maintaining the same ratio of agitator power to vessel volume does not necessarily provide the same degree of mixing — mixing a large volume is more demanding than mixing a small volume. Heat transfer is even more challenging. As vessel size increases, the volume increases more rapidly than the heat transfer area. For example, the volume of a sphere is proportional to the cube of the diameter, but the surface area is proportional to the square of the diameter. For jacketed vessels, the ratio of the heat transfer area to vessel volume decreases with vessel size.

6. Rework is a thorny issue. Suppose one of the feeds to a product batch is an intermediate. When manufacturing a product batch, a small amount of the intermediate can be off-spec material generally referred to as rework. Quality control takes a dim view of this practice, but no plant manager wants to be without it. Sometimes this is finessed with a wink and a nod, such as "the process controls will provide the capability of using rework, but we will never use this capability at this plant." Does the plant manager know if rework is being used? Sometimes he/she does; sometimes he/she does not.

These are merely examples of issues that commonly arise. But there are also items that are specific to a given product.

9.2.4. What Constitutes a Recipe

The term "recipe" as used in one part of the company can be very different from the use of that term elsewhere. The MRP system at the corporate level must know the amounts of the raw materials required to manufacture a batch of product, but parameters such as reaction temperatures are not required by the MRP system. Far more product-specific information must be provided by the recipe used at the site to manufacture a batch of product.

For the purposes of the MRP system, a "standard batch size" (such as 1000 kg of product) as in the corporate recipe is usually most convenient. Although the MRP system could potentially schedule individual batches, the MRP schedules often state manufacturing a specified amount of product. This is translated at the plant site into a number of consecutive batches (a campaign or run) that must be executed on the equipment at the plant to produce the desired quantity of final product.

Some take the position that the recipe as used within the MRP system is the "corporate" recipe. But in reality, the term "recipe" as used in the MRP context is little more the raw material list from the recipe in Table 9.1.

9.2.5. Getting a New Product into Production

For those with responsibility for the product line, shortening the time required to transition from product development to production is definitely of interest. Customers who need the new product are usually very receptive to the first company to the market with the new product. But once they start purchasing a product that fills their need, they would only consider other products if they offered a significant advantage in performance, price, or otherwise. It is easier to capture a new market than to displace a competitor from an existing market.

Someone from the development group will assist plant personnel to commission production of each new product. One activity is the translation of the product recipe from the development group to the form required by the process controls. Usually, this begins by making a copy of the most similar recipe currently used in production. Changes are then incorporated to generate the final recipe for the process controls.

One proposal for shortening the time is for the development group to use the same process controls as those used in production. Conceptually, this permits production to use the same recipe as used by the development group. If at different locations, the development group could electronically transmit the recipe to the plants! Sounds appealing, but in practice, this proves difficult to achieve and then sustain. Even if from the same manufacturer, the process controls are potentially different models, and even if the same model, different revision levels are almost certain. Furthermore, the requirements of the development group differ from requirements of production. For example, development groups need more data and higher quality data than production.

9.3. EXECUTING PRODUCT BATCHES SIMULTANEOUSLY

This issue was a major stumbling block to the application of the early programmable logic controllers (PLCs) to batch automation. The manufacturers had problems understanding that batch operations in the process industries differ from the manufacturing industries in one key aspect.

9.3.1. Separate Program for Each Product

In the 1980s, manufacturing equipment such as stamping machines were configured to make a specific part (such as the front left fender of a specific model of a car), and then a specified number of these parts were manufactured. The controller, usually a PLC, could be loaded with a program specifically for manufacturing that part. Essentially, a PLC program was developed specifically for each product, and at a point in time, only the program for that part resided in the PLC.

Translating this approach to automating a batch process meant that a program must be developed for each product; that is, for each product recipe, a corresponding PLC program would exist. The number of programs could be large, but this is not the major problem with this approach.

The introductory chapter of this book used a residential kitchen as an example of a flexible batch process. And as noted, this is not a commercial bakery—they are very different. Pies, cakes, and so on, are produced one at a time. Suppose the Thanksgiving desserts will be a cherry pie, a pecan pie, and an apple pie (the product batches). Manufacturing each pie requires a mixer and an oven (the batch units). The following must occur:

1. Mix the ingredients for the cherry pie.
2. Bake the cherry pie.
3. Mix the ingredients for the pecan pie.
4. Bake the pecan pie.
5. Mix the ingredients for the apple pie.
6. Bake the apple pie.

However, these do not occur in a purely sequential fashion. Instead, they are performed as illustrated in Figure 9.1. While the cherry pie is baking, the ingredients for the pecan pie are mixed. While the pecan pie is baking, the ingredients for the apple pie are mixed.

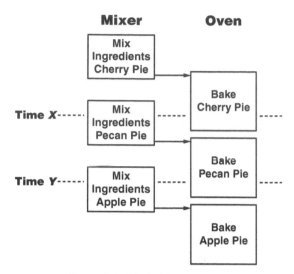

Figure 9.1. Pie baking example.

At times X and Y in Figure 9.1, two product batches are in progress. This completely nullifies the approach of preparing a program for a cherry pie, a separate program for a pecan pie, and a separate program for an apple pie. At time X in Figure 9.1, what program should be loaded in the controller? It needs both the cherry pie program and the pecan pie program. The vintage 1980 PLCs were not capable of this. Although modern controllers could perhaps provide this capability (or something comparable), other issues prevent this approach from being viable.

9.3.2. Example from the Chemical Industry

This aspect of batch processes in the chemical industry is very similar. Figure 9.2 illustrates a batch process consisting of three batch units:

Feed Tank. For each product batch, a solution must be prepared in the **Feed Tank**. Only one feed tank is illustrated in Figure 9.2, but in practice, two or more are commonly encountered.

Reactor. At some point during the batch, the solution from the **Feed Tank** is transferred to the **Reactor**. However, some activities must be completed in the **Reactor** prior to transferring the solution, and some activities must be undertaken after all of the solution has been transferred.

Neutralizer. Many reactions must be followed by further processing operations such as pH adjustment. Other examples are filtration, decanting, and so on. Basically, when all reaction activities have been completed, the contents of the **Reactor** are transferred to the **Neutralizer**. The **Neutralizer** is usually a much simpler item of equipment than the

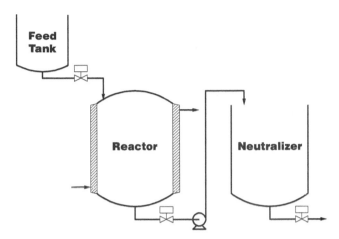

Figure 9.2. Feed tank, reactor, and neutralizer.

Reactor—no need to tie up a complex item of equipment to accomplish simple processing activities.

For this process, three product batches may be in progress at a point in time. Suppose product batch N is in progress in the **Reactor**. Once all of the solution has been transferred from the **Feed Tank** to the **Reactor**, product batch $N + 1$ can be started in the **Feed Tank**.

Activities such as pH adjustment are often not a one-shot undertaking—the neutralizing agent is added, the contents mixed, a sample taken and analyzed to determine the pH. If the pH is not within the acceptable range, the sequence is repeated. Consequently, product batch $N - 1$ could still be in the **Neutralizer** when the solution preparation for batch $N + 1$ is begun in the **Feed Tank**. If so, three product batches are in production at the same time. And in some production facilities, the three product batches could be using three different product recipes.

9.3.3. Examples from Food Processing

The equipment configurations in the food processing industry are often far more complex than those in the chemical industry. Figure 9.3 presents the configuration of a coffee roasting and packaging facility. Starting with green coffee beans, the processing is as follows:

Roast the beans. The facility in Figure 9.3 has four roasters. These are not necessarily identical, and could be some combination of batch roasters and continuous roasters.

Equilibrate. Roasting coffee beans results in an uneven moisture distribution within the beans, having more moisture in the center than near the

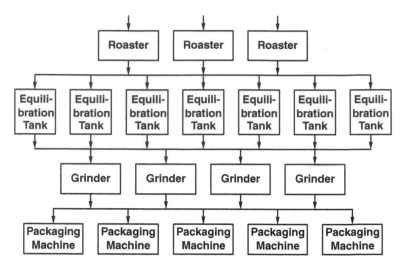

Figure 9.3. Coffee roasting, grinding, and packaging.

surface. To obtain a uniform moisture distribution within the beans, the roasted coffee beans are held within vessels known as equilibration tanks for a specified period of time.

Grind. The coffee beans are ground according to the specifications for the final product.

Package. The final product is sold in packages of various types and of various sizes.

Multiple items of each type of equipment is installed in the plant, interconnected by pneumatic conveying systems to transfer the coffee from one item of equipment to another. At a given instance of time, several product batches are in some stage of production. Pneumatic conveying systems transport coffee very rapidly, but any mistakes in positioning the diverter valves cause the coffee to very rapidly go to the wrong place.

In the coffee plant in Figure 9.3, the equilibration tanks are in fixed locations. This is not always the case in food processing plants. For example, in the manufacture of ice cream, the product from one stage of the production process is discharged to a vessel on rollers. The vessel is then physically moved to the inlet of the processing equipment for the next stage. The result is great flexibility, but with numerous opportunities for errors.

9.4. MANAGING PRODUCT BATCHES

Especially in plants where several product batches may be in some stage of manufacture at a given time, clear and explicit communications with operations personnel is essential.

9.4.1. Duration of a Product Batch

The first step in manufacturing a batch of product is to "open" a product batch, which effectively makes the product batch "known" to the process controls. Each batch is assigned a unique identifier that is used to associate operations, phases, and so on, with the product batch. The product batch remains known to the process controls until the batch is closed.

If the product recipe contains a single operation (such as simple material blending), the following approach is possible:

1. The operation could be started when the product batch is opened.
2. The product batch could be closed when the operation completes.

For recipes that contain multiple operations, complexities generally arise. For example, manufacture of one product batch may begin with operation A, but the next product batch of the same material may begin with operation B. Such variation could result from the availability of the process equipment. These and other considerations favor the following approach:

1. Opening a product batch that will be started at some time in the future.
2. Closing a product batch at a time later than when all operations have completed.

9.4.2. Making a Product Batch

A batch of product can be manufactured in at least the following three ways:

Automatic. Some use the term "recipe-driven" to refer to this mode of operation. A product recipe must be designated at the outset of the batch, and all activities are driven by specifications in the product recipe. The recipe consists of one or more operations, which in turn consist of one or more phases. Once an operation is started, the succession of phases to be executed is that specified in the product recipe.

Semiautomatic. Instead of directing activities based on a product recipe, operations personnel direct activities, usually by specifying the phases to be executed on each batch unit. In the simplest manifestation, operations personnel specify a phase to be executed on a batch unit. When this phase completes, the process controls wait until operations personnel specify the next phase to execute. More capable systems permit operations personnel to "queue up" the phases to be executed on the batch unit, which permits the process controls to proceed immediately from phase to phase until the queue is empty. So that production data can be captured when using this mode of operation, a product batch is usually opened, but no product recipe is designated.

Manual. Operations personnel direct the actions to be performed at the device level. The degree of automation, if any, is relatively low, consisting of capabilities such as automatically terminating a feed when the value from the flow totalizer attains a specified value. Phases are either not available or not used. For this mode of operation, a product batch is not opened.

The objective of automation efforts in most batch facilities is the automatic or recipe-driven mode of operation. However, the semiautomatic mode may be used initially, and thereafter for special or experimental batches.

9.4.3. Working Recipe

Some distinguish a "working recipe" from the "master recipe." When the batch is opened, the working recipe is created by copying the master recipe. This has the following implications:

1. Changes to the master recipe have no effect on any product batch currently in progress. That is, changes to the master recipe are not generally propagated to the working recipes for currently open product batches.
2. Usually the ability is provided to modify the working recipe. This enables a plant to customize a product to better meet the requirements of a customer. The product development groups can also take advantage of such a capability. However, modifications to the working recipe do not directly affect the master recipe, but generally, the modified version of the working recipe can be saved for subsequent use.

9.4.4. Opening a Product Batch

When operating in the automatic mode, several possibilities exist for opening product batches:

Corporate MRP system. Potentially, the schedule generated by the MRP system can be downloaded into the process controls to generate a schedule for the product batches. But in most plants, additional details must be resolved to translate the MRP schedule into a list of product batches. For example, the schedule from the MRP system may specify the product batches to be manufactured over some time frame (such as a week), but the following is at the discretion of the plant:
1. The order in which the batches are manufactured
2. The specific plant equipment used to manufacture each batch. In facilities such as the coffee roasting plant in Figure 9.3, the number of possible options is enormous.

Production scheduler. Most plants assign the responsibility for scheduling production to an individual, his or her main responsibility being to translate the schedule from the corporate MRP system into a list specifying the order in which product batches are to be executed on each major item of equipment. Once the list has been prepared, the production scheduler could open the appropriate product batches for the process controls. The process operators would subsequently "start" a product batch by selecting one of the batches previously opened by the scheduler.

Process operators. The list of product batches prepared by the production scheduler is provided to the process operators, who are instructed to manufacture the batches in the specified order using the specified equipment. Usually the process operators open a product batch and then immediately start it.

In deciding which approach is best, a major consideration is how to get the highest productivity from the plant, which usually focuses attention on the role of the production scheduler. In plants where hours are required to manufacture a product batch, opening a product batch is a menial task requiring little effort. The critical issue is which product batch to open to get the most from the plant.

9.5. EXECUTING OPERATIONS

Consider a chemical facility such as in Figure 9.2, except that multiple feed tanks are required. The product recipe likely contains the following operations:

1. An operation for each feed tank
2. An operation for the reactor
3. An operation for the neutralizer.

Such a recipe provides an operation for each batch unit used to manufacture the product batch. Recipes of this type are common, even though the concept of operation is not this restrictive.

9.5.1. Activating an Operation

An operation can be activated in the following ways:

By production control. As soon as a batch unit is free, production control could start an operation for a product batch. Simple logic such as "if free, start an operation" may be satisfactory for short batches using simple recipes (such as mixing materials), but for more complex recipes, other issues often arise. For the above example, production control would start operations for the reactor and for each feed tank. For the reactor, the operation for the next product batch should be started as soon as possible when the reactor becomes free. However, rarely is this appropriate for the operations on the feed tanks.

By the process operator or the shift supervisor. Some time is lost between product batches, but usually this is only of concern for the major items of equipment such as a reactor. For short-duration batches, the lost time could be significant, but for batches whose duration is measured in hours, the lost time is small relative to the total time required to manufacture a batch. Shift supervisors and process operators can base their decisions on a variety of issues, some of which may not be known to the process controls. Defining all of the issues they take into consideration is an ambitious undertaking but is necessary before this function can be automated.

From another operation. For example, the operation for the reactor could start the operation for the neutralizer when the following are true:

(a) The reactor contents are ready to be transferred to the neutralizer.

(b) The neutralizer is free; that is, no other operation is active on the neutralizer.

(c) The neutralizer is assigned or allocated to the product batch (only if equipment allocation is in use).

9.5.2. State of an Operation

Figure 9.4 presents the possible states for the operation executing on a batch unit as well as the possible transitions between the various states. The

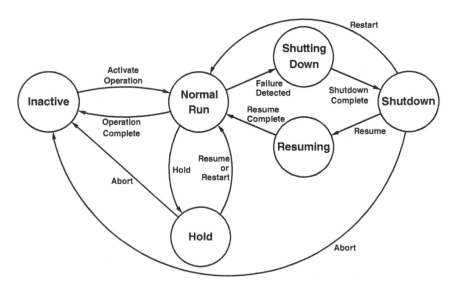

Figure 9.4. State diagram for an operation executing on a batch unit.

following four states are sustainable (in the sense that the states can exist for an extended period of time):

Inactive or Idle. No operation is active on the batch unit. The batch unit is free, and remains free until an operation is started on the batch unit.

Normal Run. An operation is active on the batch unit, and the logic is executing as expected, meaning the normal logic is being executed for the currently active phase.

Hold. The logic executing on the batch unit is not proceeding. In some situations, this state is not viable. For a train speeding down the tracks, "hold" would mean stop taking further control actions, but leave the train running. The consequences would likely break the monotony of the day. More on this shortly.

Shutdown. The batch unit has been driven to an appropriate shutdown state, which also means that the logic executing on the batch unit is not proceeding.

In the state diagram in Figure 9.4, the following two states are transient:

Shutting down. Following the detection of a failure condition, actions must be performed to place the process equipment in an appropriate **Shutdown** state. In some cases, these actions consist only of driving the discrete devices to appropriate states. For example, on detecting a high-level condition, a sufficient response may be to close the block valves on all inlet streams. But when a sequence of actions is required, the **Shutting Down** state is entered. Although similar in many respects to **Normal Run**, one important difference is that suspending or interrupting the **Shutting Down** logic is usually not advisable.

Resuming. When a sequence of actions must be performed to transition to the **Shutdown** state, a sequence of actions is usually required to restore the process to the conditions required for resuming execution of the normal logic. Sometimes this requires logic specifically to perform the required actions. The **Resuming** state is provided for executing this logic. Sometimes the normal logic can be resumed at a different point in the logic for the phase, and if so, the normal logic can be resumed directly, but at a point different from that being executed when the failure was detected.

All of this must be provided by the logic that comprises a phase. On a failure, execution of the normal logic is suspended, but in a way that it can be continued at some time in the future. To do this, the process controls must retain sufficient information regarding the currently active phase.

BATCHES AND RECIPES

9.5.3. Operator Actions

The operator and/or shift supervisor interacts with an operation executing on a batch unit using the following commands:

Activate. Valid only when the current state of the logic is **Inactive.** This command is directed to an operation within the product recipe for the product batch. Execution commences with the first phase in the operation and the state changes to **Normal Run.**

Hold. Valid only when the current state of the logic is **Normal Run.** Execution of the currently active operation on the batch unit is suspended. Process operators usually have the authority to use the **Hold** command. As will be explained shortly, this command must sometimes initiate the failure logic, which initiates a transition to the **Shutdown** state. Otherwise, the state becomes **Hold.**

Resume. Valid when the current state is **Hold** or **Shutdown.** The objective is to resume execution of the operation from the point that it was interrupted. Some plants permit operators to use this command; some restrict the command to the shift supervisor.

Restart. Valid when the current state is **Hold** or **Shutdown.** Execution of the logic proceeds from a point other than where it was interrupted. Normally, restart is only possible at the beginning of certain phases within the operation. The restart may occur at a phase prior to the one executing when interrupted, which means that some of the logic within the operation is repeated. Alternatively, the restart may occur at a later phase within the operation, which means some of the logic is skipped (presumably the skipped actions have been performed manually prior to the restart). If not used properly, the **Restart** command can have undesirable consequences. Most plants restrict this command to the shift supervisor. Even so, if an operation can be restarted at a phase, it cannot be assumed that the prior actions specified in the recipe have actually been accomplished. To the extent possible, logic should be added to the start of the phase to provide the following:

(a) Make sure the discrete devices are in their desired states, controllers are in the desired modes, and so on.

(b) To the extent possible, check that conditions within the process are as expected.

Abort. Valid when the current state is **Hold** or **Shutdown.** The command is normally restricted to the shift supervisor. If the current state is **Normal Run,** the **Hold** command must be used to transition to either the **Hold** or the **Shutdown** state. Subsequent to an **Abort,** execution of the operation cannot be recovered. The operation must either be completed manually or the **Activate** command used to begin from the first phase.

Resume and **Restart** are different in one basic respect. A **Resume** continues executing the phase that was active when the **Hold** command was issued or a failure was detected. Execution may resume at a different point within the phase, but the objective is for the phase to accomplish its purpose. A **Restart** does the following:

1. It terminates execution of the current phase.
2. It directs execution to a phase within the current operation. This may be the same phase whose execution was interrupted, but on a restart, the phase is executed from the very beginning.

A **Restart** does not expect the currently active phase to accomplish its purpose. Even if the **Restart** is at the same phase, the currently active phase is terminated before execution of the phase begins anew from the very start.

9.5.4. Consequences of the Hold Command

The **Hold** command deserves further explanation. The basic objective is to suspend execution of the phase that is currently active on a batch unit. There are situations where executing control logic can be simply suspended, but there are also situations where this is unacceptable. Consider driving an auto. Suspending control logic means remove your hands from the steering wheel — not a good idea, unless the vehicle is stopped. In this context, the **Hold** command must be understood to first stop the car and then remove your hands from the steering wheel.

Consider co-feeds to a reactor. When the co-feeds are in progress, **Hold** cannot be interpreted to stop executing the logic and leave the co-feeds flowing. Instead, **Hold** must be understood as stop the co-feeds and then suspend executing the phase logic.

For the co-feeds example, the co-feeds must be stopped on any failure condition, such as loss of a transfer pump, loss of agitation, and so on. The response to such failures involves stopping the co-feeds (and possibly some other activities). The same response is appropriate on occurrence of a **Hold** command. In effect, **Hold** is treated as an operator-initiated failure. However, this must be optional — the need to treat **Hold** as a failure depends on what activities are currently in progress on the batch unit.

9.6. BATCH HISTORY DATA

When digital systems were initially applied to automating batch facilities, plant management insisted that sufficient conventional equipment be installed such that the process operators could operate the facility independent of the process controls. This approach was basically a carryover from the approach used in continuous processes.

9.6.1. Issues with Backup Equipment

In batch facilities, provision of backup equipment led to the following experiences:

1. The reliability of the process controls was far better than feared, which meant that the process operators were not using the backup equipment on a routine basis. As is typical of infrequently performed manual operations, the error rate was high, with "making a mess" a common result. One suggestion for addressing this problem was to require that the process operators occasionally manufacture a batch using the backup equipment. Even this proved unrealistic.

2. The quality of the product suffered when operating with the backup equipment. If nothing else, process controls are consistent—they will do it the same way every time, even if it is the wrong way. Obviously, the consistent errors must be corrected, but the logic used by the process controls can be "fine-tuned" to the characteristics of the process and the equipment. The result is more consistent product with a higher quality.

3. The historical data captured by the process controls are useful in a variety of ways. Regulatory agencies such as the Food and Drug Administration (FDA) impose this requirement in certain industries, but even in the absence of this, historical data collection quickly became an essential requirement of the process controls. These data are very useful in efforts to correct errors in the logic and to "fine-tune" the logic. Despite the best efforts of QC, problems surface in the field, sometimes with the product itself but often with how the customer is using the product. Historical data on how the batch of product was manufactured are useful for both, that is, resolving real problems with the product and defending the product when the problem is on the customer's end.

The issue with backup equipment in a batch facility is fundamentally different from a continuous process. The cost of shutting down and restarting a continuous process can be noticeable, which could justify significant expenditures on backup equipment. But in a batch facility, the process (or at least a portion of it) is started up and shut down with each batch. For most batch facilities, a more realistic statement is that the process can be shut down, but it must not be down for very long.

Most plants install limited, if any, conventional backup equipment. The current generation of process controls addresses the backup issue with redundant digital controls. We seem to be able to find some very creative (and stupid) ways to shut both down, but the occurrences are very infrequent.

9.6.2. Data to be Collected

The data to be captured in a batch facility include the following:

Data values. The values of key variables are captured on a specified time interval. The storage capacity of digital equipment continues to increase, so data can be captured more frequently for more process variables. In a continuous facility, the data values are largely numerical values, with few, if any, discrete values. In a batch facility, capture of discrete values (such as agitator running/stopped) is an essential component.

Operator actions. Any action by the process operator that affects process operations should be captured. This includes starting/stopping equipment, switching controller modes, changing set points, and so on. In a batch facility, this must include any action that affects the state of the control logic executing on a batch unit, including hold, restart, and so on.

Events. The major component is alarm events, both occurrence and return-to-normal. But in a batch facility, events such as the following are often generated:

1. Events associated with the execution of the batch logic, such as operation start/end, phase start/end, and so on
2. Events associated with important items of equipment. When a process operator starts or stops an agitator, the operator action is captured. But to capture starting or stopping the agitator by the control logic executing on a batch unit, an event must be generated that the historian can capture.

9.6.3. Retrieving Data for a Product Batch

Each product batch is assigned a unique batch ID that identifies that batch of product. Knowing only this batch ID, one must be able to retrieve the data for this specific product batch. There are two approaches:

Collect and then extract. In continuous facilities, the data values, operator actions, events, and so on, are captured on a continuous basis. Applying the same technology to a batch facility to capture the data has an appeal. Retrieving the data for a specific product batch requires sorting that proceeds as follows:

1. From the batch ID, determine the batch units and the time intervals for which operations for this product batch were active.
2. Designate the key variables of interest for each batch unit (which could be part of the configuration data for the batch units).
3. Retrieve the data values, events, and operator actions for these variables over the time interval that each operation was active on the respective batch unit.

Extract and then collect. For each product batch, the objective is to generate a data file (or possibly files) containing all available information for that product batch. For the key variables for each batch unit, the data values, the operator actions, and the events are captured. One approach is to identify the key variables as part of the batch unit configuration. Another approach is to provide data collection specifications for each product recipe.

This is not an either-or choice. Some requirements for continuous data collection exist in a batch facility, usually associated with utilities (steam generation, cooling water, etc.), storage facilities, and so on.

9.6.4. Special Considerations for Batch

In some batch facilities, requirements arise for which there is no counterpart in continuous facilities. The process in Figure 9.2 is reaction followed the neutralization. The neutralization proceeds as follows:

1. Take a sample and analyze for pH.
2. If the pH is outside specified limits, then
   (a) Add a quantity of acid or base (the amount usually depends on the value of the pH).
   (b) Agitate for a specified time.
   (c) Repeat from step 1.

Collecting these data must reflect the following realities:

1. The pH analysis is not performed on a fixed time interval.
2. The number of pH analyses that are required is not known in advance.

The analysis may be performed in two ways:

By the QC laboratory. Most QC labs now have data capture facilities (often as part of a lab automation package). For QC analyses of interest to the process operators, two options are possible:
1. QC conveys the information of interest to the control room operators.
2. The control room operators have access to the data capture facility used by QC.

By the process operators in the control room. The data capture facility for QC can be extended to the control room, but often the analytical results are entered manually by the process operators.

Sometimes, the operators repeat the addition of acid or base until an acceptable pH as indicated by the analysis performed in the control room is obtained, which must then be confirmed by an analysis by QC.

A surprising number of issues can arise with regard to analyses performed on samples. The following are a few issues:

1. Sometimes, two or more analyses are performed on the same sample. Analyzers such as chromatographs provide multiple values for each analysis.
2. "Time" has several possible interpretations, all of which could potentially be of interest:
    (a) The time the sample was taken. Some further distinguish between the time the sample was scheduled to be taken (the "log time") and the time the sample was actually taken. For example, the sample scheduled to be taken at 10:30 (the log time) was actually taken at 10:23. Searching the database by the log time is simpler than searching by actual time.
    (b) The time the analysis was performed.
    (c) The time the data value was entered into the system.
3. Samples may be missing. Perhaps the sample was never taken; perhaps the sample was subsequently "lost" in some manner (dropped, contaminated, etc.).
4. Under certain situations, some or all analyses performed on a sample can be repeated. The simplest approach is to replace the original analysis with the subsequent analysis, but more likely all analyses must be retained.

In the end, the complexity far exceeds what is usually anticipated. Most lab automation systems address such issues, which makes their use appealing.

9.7. PERFORMANCE PARAMETERS

These fall into two categories:

Parameters relating to product quality. ISO certifications require that key process variables be identified and limits established for each.

Parameters relating to plant performance. These performance parameters refer to quantities computed with the specific objective of assessing the performance of a facility. This always raises issues, some of which have overtones in corporate politics.

Only the latter category of performance parameters is the subject of this section.

9.7.1. Product Profit Issues

Ultimately, the question is economics, but even this is not entirely cut and dried. In a flexible batch facility, a variety of products are manufactured. The profit margins are higher for some products than others. All plant managers want to manufacture those products with the high profit margins — plants manufacturing mostly products with high profit margins will look better economically than those manufacturing the products with low profit margins.

But from the corporate perspective, manufacturing a product with a low profit margin is preferable to an idle plant, so somebody will have to make them. Just because a plant is generating the most profits does not mean that the plant is being operated more effectively than another plant; it may just have a slate of products with higher profit margins.

9.7.2. Computing a Performance Parameter

The way a performance parameter is computed sometimes depends on how that performance parameter will be used. Unfortunately, such discussions have political overtones, so a contrived example will be used to illustrate the point.

Suppose plant productivity will be computed as the amount of product per hour, the units being kilogram/hour and computed over a 1-week interval with agreed upon starting and ending times. The plant runs 24 hours a day, 7 days a week. The number is simply computed as the total kilogram of product produced for the week divided by 168 hours.

Now for the complication. The plant operated very successfully for several years. But then, an eastern religious sect chose the town for their new home. Their members were avid voters and proceeded to take over the town council. They deemed it appropriate to spend the time between noon and 2:00 PM every Thursday in prayer, and established an ordinance to this effect. Being a liberal sect, the ordinance permits one to pray to whatever they like — even a door knob suffices! But one cannot operate a plant while praying to a door knob. So every week beginning at noon on Thursday, the plant is shutdown for 2 hours.

Does this impact the way plant productivity should be computed? In most cases, certainly not. If one is deciding where to invest in new production facilities, the lower productivity for this plant makes it the inappropriate choice.

Management likes to create spreadsheets for various purposes. Suppose the performance of the plant manager is being assessed. Envision a spreadsheet with the names of plant managers on the columns and various performance parameters on the rows. One of the performance parameters is the plant productivity. Is the plant productivity as computed above appropriate for this spreadsheet? The plant is shut down for 2 hours each week, but there is nothing the plant manager can do about it. For comparing the performance of the various plant managers, the plant productivity is the amount of product produced divided by the hours that the plant is available to be run. The amount of product produced should be divided by 166 hours, not by 168 hours.

Such discussions can become very heated. Whenever possible, each plant wants a performance parameter computed in a way that makes that plant look better.

9.7.3. Batch Times

Even seemingly simple quantities like batch times have to be carefully defined. Consider the reaction-neutralization facility in Figure 9.2. How does one define the time required to produce a batch of product? Most batches start with the preparation of the solution in the feed tank, although it is possible for a batch to start with charging the heel to the reactor. All batches end when the neutralization is accomplished. Consequently, the batch time could be computed from the following:

Start time. The earliest of starting the solution preparation in the feed tank and charging the heel to the reactor.

End time. The completion of the neutralization.

The batch time is the difference between these two.

One approach to representing and analyzing production activities in a batch facility is to apply tools from project management, such as the critical path method (CPM). The starting point is to define all tasks or activities, and for each, specify the time duration and the dependencies (i.e., tasks that must be completed before a given task can commence). Various graphical representations used in project management can be adapted for use in batch.

For the batch process in Figure 9.2, the tasks might be as follows:

Batch Unit	Activity	Duration (h)	Dependency
Feed Tank	Solution Prep	2.25	
	Co-Feed	4.0	Solution Prep
Reactor	Charge Heel	0.75	
	Co-Feed	4.0	Charge Heel
	React	1.5	Co-Feed
	Transfer	0.75	React
Neutralizer	Transfer	0.75	
	Adjust pH	2.0	Transfer

The tasks are listed separately for each batch unit. However, tasks such as **Co-Feed** and **Transfer** in the aforementioned list involve multiple batch units. Although duplicated in the above list, each entry must have the same duration time and cannot commence until all dependencies are satisfied. For task **Co-Feed** to commence, task **Solution Prep** must be completed in the **Feed Tank**, and task **Charge Heel** must be completed in the **Reactor**. The above example is for a simple batch, so tasks are executed in a linear fashion on each batch unit. However, more complex structures with alternate paths and the like are possible.

Figure 9.5 uses a variation of the Gantt chart to illustrate the sequence of tasks required to manufacture a product batch. The tasks are arranged so as to manufacture the batch in the manner that gives the shortest possible value

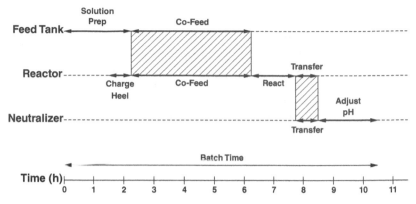

Figure 9.5. Tasks for a single product batch.

of the batch time as per the definition presented previously. The following observations can be made:

1. The shortest possible batch time is 10.5 hours.
2. The **Charge Heel** task on the **Reactor** could be started 1.5 hours earlier without having any effect on the batch time.

However, there are problems with both of these statements, specifically, the following:

1. Attaining the shortest batch time as per the above definition is not a smart manufacturing strategy.
2. Starting the **Charge Heel** task at the same time as the **Solution Prep** task in the **Feed Tank** will result in 1.5 hours idle time in the **Reactor**, which is usually not desirable.

The deficiency with the above analysis is that it examines only one product batch. In practice, a series of product batches is manufactured. Figure 9.6 is similar to Figure 9.5 except that it illustrates a series of product batches that manufacture the same product. As illustrated in Figure 9.6, the batch time is still 10.5 hours. However, this assumes the following:

1. **Solution Prep** is begun in the **Feed Tank** at the latest possible time, which means it completes at the same time as **Charge Heel** in the **Reactor**.
2. **Upon** completion of the transfer from the **Reactor, Adjust pH** is promptly undertaken in the **Neutralizer** and completed as expeditiously as possible.

Neither of these is necessary and could potentially be counterproductive.

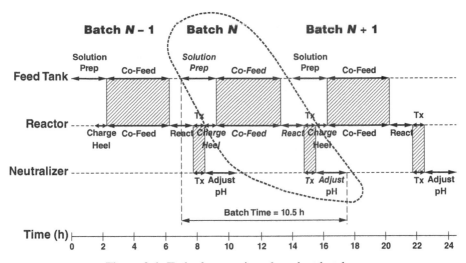

Figure 9.6. Tasks for a series of product batches.

In project management, the objective of the critical path method is to identify those tasks or activities that are on the "critical path," meaning that a delay in any of these activities would extend the time required to complete the job. Figure 9.6 indicates that all activities in the **Reactor** are on the critical path. Any delay in any of these activities has a negative impact on plant productivity. Activities off the critical path must be completed so as to not delay any activity in the **Reactor**. There will be a latest possible starting time, but starting the activity earlier is certainly acceptable and often appropriate.

First, consider **Solution Prep**. So as to not delay activities in the **Reactor**, **Solution Prep** must complete prior to the completion of **Charge Heel** in the **Reactor**, but there are no downsides to completing **Solution Prep** earlier. In the chart in Figure 9.7, **Solution Prep** for **Batch N** starts as soon as **Co-Feed** for the previous batch completes. This results in an idle time of 45 minutes in **Feed Tank** following completion of **Solution Prep**. In Figure 9.6, this idle time preceded **Solution Prep**. Whether the idle time is prior to or subsequent to **Solution Prep** is immaterial.

In practice, **Solution Prep** should be started at the earliest time. Occasionally, problems surface during **Solution Prep**. Starting **Solution Prep** at the earliest possible time provides 45 minutes to resolve such problems before they delay activities in the **Reactor**.

Starting **Solution Prep** at the earliest time increases the batch time (as defined earlier) from 10.5 hours to 11.25 hours. If the focus is simply to minimize the batch time, the appropriate behavior for the process operators would be to start **Solution Prep** at the latest possible time. But should any problems surface during **Solution Prep**, the result is a delay in the activities in the **Reactor**. **Solution Prep** should be started at the earliest possible time, not at the latest possible time.

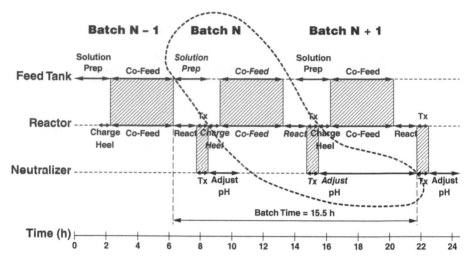

Figure 9.7. Alternate execution of tasks for a series of product batches.

Now consider **Adjust pH**. If **Adjust pH** is started at the first opportunity and completed expeditiously, **Adjust pH** is followed by an idle time of 4.25 hours in the **Neutralizer**. Starting **Adjust pH** could be delayed for 4.25 hours with no impact on plant productivity. However, a more prudent approach would be to promptly proceed with **Adjust pH**, but with an "as time permits" attitude. Specifically, any demand from the **Reactor** for the operator's attention should take priority over any demand from the **Neutralizer**. Provided **Adjust pH** completes within 6.25 hours or less, no activity in the **Reactor** will be delayed.

In Figure 9.7, the time for **Adjust pH** is 6.25 hours. Also, **Solution Prep** is started at the first opportunity. As computed by the definition presented previously, the batch time is 15.5 hours. In Figure 9.6, the batch time is 10.5 hours. However, the plant productivity is the same for Figure 9.6 and Figure 9.7. Just because the batch time (as defined earlier) is longer does not necessarily mean that the plant productivity is less.

If a time-based parameter is to provide an effective assessment of plant productivity, the focus must be on the tasks on the critical path. The problem with batch time as defined previously is that its start time is for a task (**Solution Prep**) not on the critical path, and its end time is also for a task (**Adjust pH**) not on the critical path.

To obtain a true assessment of plant productivity, one possibility is to use the time between the completion of **Transfer** for the previous batch to the completion of **Transfer** for the current batch. The expected value is 7.0 hours. Any increase in this value translates to a decrease in productivity.

INDEX

Abort, 308
Accuracy, 37, 41, 49
Acquire, *see* Enqueue
Action, 223
Activate, 308
Agitator seal, 46
Alarm, 251, 266, 271, 311
Analog I/O, 149
Analog values, 26
Analysis time, 313
Atmospheric, 40, 234
Auctioneer, *see* Selector block
Automatic, 303
Automatic tuners, 102
Averaging time, 55

Barcodes, 204
Batch ID, 214
Batch process, 3
Batch size, 213
Batch time, 315
Batch unit, 227
Bleed back, 63
Blowing steam, 123

Broken wire, 151
Bypass, 138, 145

Calibration, 36
Cascade control, 18, 28, 114, 118, 131
Cavitation, 139
Characterization function, 29, 48, 100, 135, 146
Check meter, 63, 197
Chemical name, 215
Co-feed, 85, 231, 288
Common name, 215
Compiler, 294
Compression, 49
Condensate pot, 124
Continuity, 151
Continuous control, 224
Continuous process, 3
Control logic diagram, 78
Controlled flow, 95
Controlled outputs, 250
Controller gain, 103
Controller tuning, 102
Convergent AND, 286

Control of Batch Processes, First Edition. Cecil L. Smith.
© 2014 John Wiley & Sons, Inc. Published 2014 by John Wiley & Sons, Inc.